Science and *Con*science

New York COLUMBIA UNIVERSITY PRESS *1980*

Library of Congress Cataloging in Publication Data

Wessel, Milton R
Science and conscience.

Includes bibliographical references and index.
1. Science—Social aspects. I. Title.
Q175.5.W47 303.4′83 80-12780
ISBN 0-231-04746-0

Columbia University Press
New York Guildford, Surrey

To my colleagues in
law, science, and
business

CONTENTS

PREFACE

This book is about the modern "socioscientific dispute"—the complex, public-interest controversy which deals with human "quality-of-life" problems. Examples, at the time of this writing, are the disputes over nuclear power (Three Mile Island and Seabrook), toxic compounds ("Agent Orange" and saccharin), transportation safety (the DC-10 disaster), concentration of economic power (the decade-long IBM suit), "reverse discrimination" (the *Bakke* case and its progeny), inflation control, energy conservation and alternatives, and insuring Russian compliance with the SALT treaties. Things won't stand still for our study, however. It is a safe bet that there will be new and different issues boiling when this book first appears in print some months hence, and later whenever it is being read. The problems are growing, just like Topsy.

We are all lay persons, including even the scientist outside of his special discipline. We are moving further and further into a high-science, high-technology, information-based society, most parts of which are far beyond our individual competence. The signs are clearly all around us: computers; nuclear weapons; space travel; instant, worldwide communications orally, visually, and by document; heart and other vital-organ transplants; laser operations; microsurgical attachments; X ray three-dimensional scanning; "cloning"; genetic manipulation; and *so* much else. It is no wonder that many people have the uneasy feeling, "It's beyond me," and "I can't cope." President Carter in mid-1979 spoke of a "crisis of confidence" in the country. The problem he correctly identified, however, is not one created by inadequate leadership alone, or any other single, simple cause. It will not go away as the result of any single, simple solution by him, such as a shake-up of the U.S. Cabinet. It is a problem which first began to emerge when the atom bomb was dropped on Hiroshima on August 6, 1945. It has been developing ever since.

As we enter the ninth decade of the twentieth century, science and technology are far outrunning and outdistancing the society they are supposed to serve. The resulting concerns are burgeoning. The legal profession, corporate society, science, and even government are all losing credibility in the process. Large segments of the general public no longer know whom to trust, and what to believe. There is danger that, in its despair, the public will turn to ever-greater centralized government regulation and control, with loss of individual freedom as the inevitable result.

These concerns will begin to decline only when society learns how to manage and control science and technology in its own best interests. The societal institutions and mechanisms of an earlier day, designed for earlier problems and concerns, must be improved or changed so as to match modern developments. This is an enormous task, which cuts across the length and breadth of society itself. This book deals with one aspect of the assignment, the resolution of disputes about scientific, public-interest matters.

Socioscientific disputes have a number of important special characteristics which distinguish them from older, more traditional disputes. These features make it difficult for our existing, traditional, adversarial dispute-resolution mechanisms to resolve them satisfactorily in the public interest. One important result has been an increase in the extremes of "adversarialism." This has put more and more people and organizations at each other's throats. Bad, and sometimes unacceptable, decisions in socioscientific controversies are often the unfortunate consequence.

The first task is for all of us to examine where we are and how we got here. This will help us to understand what is wrong, and to begin to develop solutions. Those solutions must be new and creative ones, geared to computers and nuclear power, not the horse and buggy. Our world is changing rapidly and dramatically. It is not the answer for lawyers just to revise and re-revise their canons of ethics and to redoctor the discovery rules; or for industry to "step-up" its public information activities and teach people the value of economic freedom; or for science to intensify its search for more and better information. Of course, we need all of these, but we need much, much more. When we stop and look at where we are, it becomes apparent that we are using Middle Ages tools to work with Space Age problems. We need data-gathering, analysis, discussion, debate, writing, teaching—all the things people do to understand and solve. Most of all, we need "imagineering." When we have enough of all these, we will discover how to make science "work" for us all.

This book is not simply a plea for study and analysis, however. I believe that some of the things we must do are already well defined and ready for intensified testing and application. I believe that law, business, and science must each get more public-interest "responsibility" into its discipline. I am convinced that the "rule of reason" should be the guiding philosophy employed in socioscientific dispute resolution. I am urging, finally, that science modernize elements of its traditional "scientific method." That will make it possible to reach a new interim scientific consensus. It will enable dispute resolution to factor more valid scientific findings into its conclusions.

I do not advance these proposals either as cure-alls or complete answers. I have hardly discussed the role of government. My personal view is that the private sector should develop its own answers if it can. It should turn to government only as a last resort. We have not really even begun to examine the problem, much less test new mechanisms. Bad as the "crisis" may be, it is not yet sufficiently serious to justify governmentally imposed restraints on freedom. I appreciate, of course, that others will differ with me on this and many other points. My purpose will have been served if others use my proposals only as points of departure. At least we will then be on our way to better dispute resolution.

The structure of this book is, first, to analyze our dispute-resolution problems, and then to propose solutions. It breaks down as follows:

—Chapter 1 analyzes the differences between the modern socioscientific and the traditional dispute. It discusses the impact of science on society. It shows how technological developments since World War II have brought us to the present "risk/benefit" period of dispute resolution.

—Chapter 2 treats the extremes of "adversarialism" which permeate our society. They are primary root causes of our dispute-resolution problems.

—Chapter 3 deals with the changes which have taken place in the legal profession and the practice of law since World War II. It asks for greater "legal responsibility" in business-law practice.

—Chapter 4 treats the changes in corporate society since World War II. It identifies corporate responsibility, through long-run profit maximization, as the corporate public-interest ethic which will improve dispute resolution.

—Chapter 5 explores the lawyer-corporation relationship. It offers several specific suggestions for changes in the management of the corporate legal function.

—Chapter 6 discusses "adversary science." It suggests a number of ways by which science can communicate better with the lay public. It pleads for a science public-interest ethic, by way of "science responsibility."

—Chapter 7 proposes the scientific "consensus-finding conference" as a new mechanism to deal with socioscientific problems.

—Chapter 8 examines the acrimony which has come to characterize so many socioscientific disputes. It seeks an end to the vilification and hatred which stand in the way of finding solutions in the public interest.

Intended Audience

The first thing any author should do, both for himself as writer and for those who must decide whether to read, is to identify his target audience. This tells him how to write, and the reader—the sole object of any book—whether to read.

I have struggled with the question of target audience ever since the plan for this book first formed in my mind. So have persons whose opinions and complete critical honesty with me I value most highly: My wife, Joan, who has helped write every book I have written; Bruce Gilchrist, a computer scientist and valued colleague of many years, who was coauthor of the first book in this series, begun almost a decade ago; and Will Eisner, the noted illustrator who helped so much to communicate the concepts of the second book in the series, with his pointed and very dramatic illustrations. As each reviewed draft after draft, each repeatedly asked me, "Just who *is* your audience?" Will struggled mightily to design useful illustrations. He was the first to concede that he could not achieve our mutual purpose of enhanced communication through additional sensory perception. He could not clarify a complex concept by single-glance picture or through the "sequential art" in which he has pioneered. The fault was not his. It was mine. I have not been able to furnish a satisfactory answer to his question, so that he could draw for a specific and limited audience.

This book deals with many highly specialized, esoteric disciplines. The principal roles in the disputes involved are played by lawyers, the senior managers of large corporations, scientists, the leaders of activist groups, and government officials. These groups are, of course, all important parts of my target audience.

Other books have dealt with problems involving several such disciplines, however, and the authors seem to have had no great difficulty identifying and speaking to the persons they sought to persuade. The trouble with *this* book is that the concerned "public" at large plays the ultimate and truly significant role in the disputes under discussion. The primary target audience of this book includes not only the lawyer, the industrialist, and the scientist—all of whose credibilities have been so impaired in recent years; and the congressperson, and the Treasury Department and Federal Reserve Board executive dealing with inflation; and the diplomat and military official involved in SALT enforcement—all of whom deal directly with socioscientific dispute resolution. It also includes the consumer, who smokes, uses saccharin, takes medications with side effects, and flys in a DC-10. It includes the farmer, who applies a pesticide. It includes the family which lives near Three Mile Island or the Love Canal disposal area. It includes the computer software vendor, who worries about IBM's next product. It includes the activist, who resists reverse discrimination. It includes the next generation—my children, my students, and those others who will be running this world in a few years, and who with some justification are so cynical today. These people, and many, many others, are the ones who must ultimately be satisfied with the ways in which we, as a society, resolve the scientific disputes which determine their fates.

They are the people who, in a democracy, should be the final arbiters of important, controversial affairs. They will decide whether radical changes in our whole economic or social systems are to be made. They make up my real target audience.

My answer to Joan, Bruce, and Will, then, is, "This book is addressed to every concerned person." The result, as they have pointed out, is that this is not an easy book to read. I have struggled to make it simpler and more readable, in the hope of gaining a larger audience. My efforts, however, have been met with the far more serious criticism that I was being simplistic. This might encourage a larger lay audience, but it would lose the expert who must be the leader in societal change. I have finally decided to try no more. I apologize for the difficulty, and hope the reader will consider the importance of the subject worth the effort of more hours of work than a book this size would normally require.

Data Base

Only in very recent years has the socioscientific dispute even begun to receive recognition as posing serious problems. There is still but a very small data base with which to work. Most of the cases, which we lawyers like to cite for precedent, are now working their way through the very earliest phases of the dispute-resolution process. They are being discussed by corporate boards of directors, by government executives, and by citizen group leaderships. Most have not yet reached the administrative agency action stage, much less the later stages of agency hearing and decision, or judicial trial and appeal. With few exceptions, not a fair sampling by any measure, final U.S. Supreme Court action is many years away.

Nor have our scholars adequately addressed the issues. There are many persons, in and out of academe, who have done brilliant work in specific subjects, such as nuclear power, carcinogenicity, economics, or civil rights. Very few, however, have approached all the disciplines of "science" as one subject, together with all the mechanisms for resolving disputes, and all the elements of society which are affected. The result is that there is very little good precedent to cite—few cases, little empirical data, few books, and little analysis. More often than I would like, I have had to refer to cases and authorities which are not directly in point, and which are useful only by analogy. A good part of my references is to current events and comments, as reported in newspapers and periodicals. One trouble with this is that accuracy is sometimes suspect. Where there is doubt, I have used such sources only where the main purpose is to establish perception, not truth. Perception, as evidenced by what is published and said, can be real enough—even if the media and the public have got the whole thing backwards.

In addition to this potential for inaccuracy, in almost all cases there is another side to be heard. Some of those involved are friends, colleagues, or clients. Where anonymity has seemed appropriate and possible, therefore, I have avoided identifying personalities. This may pose a burden on some readers, but it should not interfere with logic or argument. We are dealing with societal mechanisms, not individuals. If what is going on is wrong, it is the system which is to blame, not specific people or organizations doing what the system allows them to do. The reader who needs to know more for any reason, can of course go back to the cited work itself.

I have had a great deal of help from many experts in many disciplines. The matters treated in this book have been the subject of seminars with graduate lawyers, law students, and business executives, going back to 1972. Many of the specific concepts were presented to, and criticized by, members of the faculties at Stanford Law School in 1978, and at Columbia Law School in 1979. The Columbia Law School Center for Law and Economic Studies, organized a most helpful seminar to critique some of the concepts. It has also supported the assistance provided by Judy Yavitz, a Columbia Law School student. She has shepherded the final manuscript through to completion and provided the invaluable guidance and critique of another generation and perception.

I have circulated preliminary prospectuses and summaries—and several comment drafts at various stages of the manuscript—to a large number of clients, colleagues, and friends in law, business, and science. In acknowledging their help, I must express even more than the author's usual caveat that he alone is to blame for errors. Only a few people have seen the final manuscript. None of the others knows the extent to which I have accepted criticism or suggestion. Indeed, some will find that I have voiced my disagreement with a criticism. Clearly I alone am responsible for what I have written.

I am most grateful to those persons who reviewed and commented on earlier drafts of portions of the manuscript, or who lent their encouragement or pressed me to move in new directions: Bertram A. Abrams, Earle B. Barnes, Harlan M. Blake, Thomas W. Carmody, Warren B. Crummett, Harold S. H. Edgar, Will Eisner, Walter Gellhorn, Harvey J. Goldschmid, Cleve A. I. Goring, Bernard Gronert, John Kaplan, Rick Main, James A. Martin, Francis X. Quinn, David Rosen, Monroe S. Sadler, Michael I. Sovern, and Ralph C. Wands, and Murray Dolmatch, who suggested the title.

I must acknowledge a special debt to those who, at my request and sometimes urging, reviewed and critiqued the manuscript in its later stages: R. William Barker, Richard L. Berman, Etcyl H. Blair, Jules N. Bloch, Frank M. Brower, Richard L. Crandall, Eleanor M. Fox, Bruce Gilchrist, Leon Golberg, William A. Groening, Jr., Irving Like, Herbert E. Marks, James F. Mathis, Francis X. Murray, Leslie F. Nute, Stanley D. Robinson, V. K. Rowe, I. W. Tucker, Joan S. Wessel, and Macauley Whiting.

Background and Qualifications

Bruce Gilchrist and I wrote in the preface to the first book in what is now a four-volume series (with no end in sight): "A study such as this one must be influenced to some extent by the individuals conducting it. Therefore we believe that our backgrounds are relevant." We had not realized how right we were. Publication of the book was followed almost immediately by an effort to suppress it, by a vote to censure one of us in a scientific society, and by the withdrawal of one of the guest participants in a law school seminar I was teaching.

I would like to believe that everyone will consider this a completely impartial and strictly scholarly work. But my earlier experience with Gilchrist, and similar experiences thereafter, have taught me better. The issues involved in socioscientific dispute resolution evoke strong, and even intense, emotions. Some persons in the "environmental" camp, view me as a committed and activist adversary. Others, on the "industry" side, see me as too accommodating, and even as "disloyal" and destructive of fundamental, free enterprise principles. A few people who were once colleagues, friends, and contributors to my earlier work, will no longer deal freely with me. They have been unwilling to comment on the present book. Most people consider themselves to be fair, unbiased, and unprejudiced, and I am no exception. However, that is obviously not the way in which all others see me. All of this history tells me that it is probably even more important for me to disclose my background now than when I began. The reader can then judge bias for himself. Moreover, my qualifications are, of course, relevant to an evaluation of my ability to speak accurately of the many different disciplines treated in this book.

I have been a trial lawyer since graduating from law school in 1948. I have been involved in all kinds of litigations. For the first four years of my career, I worked exclusively on *Ferguson* v. *Ford Motor Co.*, the grandfather of all massive antitrust cases. Thereafter, I worked on the international oil cartel case and several other major antitrust litigations, and the wide variety of matters to be expected in a litigation career spanning three decades. Included were negligence cases (I handled all local tort cases against the U.S. government during one year while I was an assistant U.S. Attorney), marital litigations, criminal prosecutions and defenses, and both sides of environmental matters. I was even a speaker at the Audubon Society convention which sparked the creation of the Environmental Defense Fund (EDF), although I assure my industry friends that I have no right to credit for EDF's existence!

Despite this variety, most of my practice has been industry oriented. From 1948 to 1953, 1956 to 1958, and 1960 to 1973, I was associated with two giant law firms in New York City, primarily with corporate practice. I practiced law alone from 1974 to 1979, but perhaps with even more of a business bias. I am now serving as counsel to another large New York/Washington law firm, again with a business practice.

I have no academic expertise in any of the disciplines involved in this book other than law. However, I have been involved with computer industry problems and concerns since 1964, and with the chemical industry since 1970. I have been General Counsel to the Association of Data Processing Service Organizations (ADAPSO) since 1966, and have served as corporate director of three small, public computer corporations. I was General Counsel to the American Federation of Information Processing Societies (AFIPS), the professional society federation of the computer industry, during the period 1967–1975. I have been teaching a course in "computer law" at various law schools (presently Columbia) since 1972.

I have been General Counsel to the Chemical Industry Institute of Toxicology (CIIT) since its organization in 1974. I served as Special Litigation Counsel to the Dow Chemical Company for almost five years, until June 30, 1978. I represented Dow in connection with aspects of the 2,4,5-T ("Agent Orange") and National Coal Policy Project (NCPP) issues discussed in chapter 7. I have lectured, advised, and taught seminars on dispute resolution to a number of other chemical companies.

In 1970, Bruce Gilchrist and I began the discussions, investigation, and research which led to our publication of *Government Regulation of the Computer Industry* in 1972. It was not planned that way but, as I look back, that early work has turned into what is now an almost decade-long project in "science and society." The project has included teaching of law school seminars in "socioscientific dispute resolution" at Stanford, New York University, and now Columbia law schools, and three more books.

Government Regulation dealt with a specific science-impact area—computers. It pointed out that few persons seemed to be paying any attention to the effect of computers on industry and society generally. It encouraged the kind of coordinated understanding and approach which was later reflected in the studies of the Privacy and Electronic Funds Transfer commissions, and in an increasing number of other investigations into the impact of computers on society.

Computers turned out to be a good science for study. Unlike nuclear power or chemical carcinogenicity, for example, there was relatively little animus, or even strong emotion, about computer impact problems (except for feelings about IBM, pro and con). The most serious early computer impact issue was "privacy," and this evoked the same emotions as apple pie and mother love. The result is that I have been able to work in a comparatively uncharged atmosphere insofar as the computer science aspect of this project is concerned.

My second book in this sequence was *Freedom's Edge*, published in 1974. It analyzed a number of computer societal impact issues. It urged the balancing of benefits against dangers, and extension of the "risk/benefit" approach then developing in the environmental area. The third book, *The Rule of Reason*, was published in 1976. It moved from analysis of science impact to dispute resolution. It concluded that the employment of adversarial extremes was a major cause of the breakdown of corporate dispute resolution. It pleaded for a change to the "rule of reason," as the preferred dispute-resolution philosophy.

Some readers, at first glance, may consider that what I propose in this book is naive or utopian. I appreciate that a claim of special qualification or expertise by virtue of one's own necessarily limited experience, and without the extensive empirical analysis which this kind of effort deserves, may be considered "bootstrapping." Nevertheless, I *have* been at it a long time. I hope that I have, at least, earned the right to ask that dubious readers examine the issues again, more carefully. I am convinced that my suggestions are practical, and even "self-serving," for the user (as they have been accused of being), as well as moral and ethical. In addition, perhaps a model of utopia will be useful. We could use a little more idealism among those involved in resolving disputes about matters that will determine the future of this world.

Columbia Law School
New York, N.Y.
May 1980

ABBREVIATIONS USED IN TEXT AND NOTES

ABA	American Bar Association
ABAF	American Bar Association Foundation
ABAJ	*American Bar Association Journal*
ACLU	American Civil Liberties Union
ADAPSO	Association of Data Processing Service Organizations
AFBF	American Farm Bureau Federation
AFIPS	American Federation of Information Processing Societies
AID	Agency for International Development
ATRR	*Antitrust and Trade Regulation Reports*
BNA	Bureau of National Affairs
BW	*Business Week*
C&EN	*Chemical & Engineering News*
CEA	Council of Economic Advisers
CEQ	Council on Environmental Quality
CIIT	Chemical Industry Institute of Toxicology
Con Ed	Consolidated Edison Company of New York
CPSC	Consumer Product Safety Commission
CSIS	Center for Strategic and International Studies, Georgetown University
CW	*Chemical Week*
DES	Diethylstilbestrol
DNA	Deoxyribonucleic acid
ECCS	Emergency Core Cooling System

EDF	Environmental Defense Fund
EPA	Environmental Protection Agency
ESA	Endangered Species Act
FAA	Federal Aviation Administration
FCC	Federal Communications Commission
FCPA	Federal Corrupt Practices Act
FDA	Food & Drug Administration
FIFRA	Federal Insecticide, Fungicide and Rodenticide Act
FTC	Federal Trade Commission
GSA	General Services Adminstration
HEW	Department of Health, Education, and Welfare
ICC	Interstate Commerce Commission
Legal Times	*Legal Times of Washington*
LEAA	Law Enforcement Assistance Administration
LNG	Liquified Natural Gas
NCPP	National Coal Policy Project
NCRALP	National Commission for the Review of Antitrust Laws and Procedures
NCTR	National Center for Toxicological Research
NEPA	National Environmental Policy Act
NHTSA	National Highway Traffic Safety Administration
NIEHS	National Institute of Environmental Health Sciences
NIH	National Institutes of Health
NIOSH	National Institute for Occupational Safety and Health
NLJ	*National Law Journal*
NRC	Nuclear Regulatory Commission
NRDC	Natural Resources Defense Council
NSF	National Science Foundation
NYT	*The New York Times*
OMB	Office of Manpower and Budget

OSHA	Occupational Safety & Health Administration
OSTP	Office of Science and Technology Policy
OTA	Office of Technology Assessment
OTS	Ortho-toluene sulfonamide
PCN	*Pesticide & Toxic Chemical News*
PPM	Parts per million
PPB	Parts per billion
PPT	Parts per trillion
SALT	Strategic Arms Limitation Treaty
SOT	Society of Toxicology
TCDD	2,3,7,8-tetrachlorodibenzo-para-dioxin
TSCA	Toxic Substances Control Act
2,4,5-T	2,4,5-trichlorophenoxyacetic acid
USDA	United States Department of Agriculture
USLW	*United States Law Week*
WSJ	*Wall Street Journal*

Science and *Con*science

Scientists may develop and demonstrate [nuclear] disposal technology, but the rest of us must not only decide profoundly human political questions; we must decide *how to decide*. The question is not just where to locate a waste repository but how that decision can be equitably made and how people can be brought to accept it. How can the public be satisfied as to the efficacy of the technology? What role for the states? What for localities? Can anything less than the best site for technological (safety) purposes ever be acceptable, for political or any other reasons?

Such questions can't be addressed by scientists alone. They are not being satisfactorily addressed by anyone else, in government or industry. Yet, more than any others, these questions ultimately will decide the future—if any—of nuclear power; and no issue in American life today is of more importance, to current generations and whatever posterity there may be.

—Tom Wicker, "The Real Nuclear Issue,"
"Op-Ed Page," *New York Times*, May 11, 1979

1

SCIENCE AND SOCIETY

Nuclear power offers the United States a partial escape from long gas lines, skyrockcting energy costs, and reliance on imported oil. Is it really a sound economic proposition when rigorously costed out? Does its spread increase the likelihood of nuclear war, or the theft of bomb materials by terrorists? Is it "safe" to live near a nuclear reactor? Can atomic wastes be disposed of safely for thousands of years? Can our society *afford* nuclear power?

Genetic laboratory research with deoxyribonucleic acid ("recombinant DNA research") promises to uncover many of the secrets of life, aid in the prevention and cure of cancer and other disease, and produce substances which will help control environmental hazards. Does it also threaten to create uncontrollable forms of life which may escape from the laboratory and destroy the human race—a man-created "Andromeda strain?" Can our society *afford* recombinant DNA research?

Computer technology is one of the very few resources whose effective throughput cost has dropped dramatically even during escalating inflation. Computer systems can inform instantly, improve education, limit the need for manual labor, and enhance productivity. Do they also interfere with privacy, centralize power in the hands of a few, isolate those who do not conform or are somehow "different," and otherwise infringe human freedom? Can our society *afford* computers?[1]

American industry has developed special employment, education, and training programs designed to improve the economic opportunities of members of minority groups against which our society has discriminated. Do these programs discriminate unfairly against others? Can our society *afford* "reverse discrimination?"

These are examples of the major kinds of "socioscientific dispute," a new and different type of public policy problem which has come to prominence since World War II. The list is growing. Its scope and range are already enormous, and cut broadly across society generally. There is something in it to entice almost every person who is concerned with the quality of our life on this planet. For instance:

Laetrile	Does Laetrile cause postponement of more effective medical therapies? or create a significant risk of cyanide poisoning?
DES	Does feeding diethylstilbestrol (DES) to cattle also carry a serious risk of cancer to those who eat the meat?
Saccharin	Does the use of saccharin pose an unacceptable risk of human cancer?
Peanut butter	Should the marketing of peanut butter be limited or banned, because it contains a suspect carcinogen? Should the same be true for red dye #2, or #4, or #40, all widely used colorings, or nitrite, used to control botulism and other food poisoning and improve taste?
Food preparation	Should the frying of hamburger and other meats be restricted, because this form of cooking increases the presence of carcinogens?
Ozone	Does supersonic transport threaten to impair the stratosphere's ozone layer? Do fluorocarbons do so also?
LNG	Does the use of liquified natural gas (LNG) result in an unreasonable risk of harm to human life and the environment?
Energy	Shall we accept greater air pollution from coal use in order to limit dependence on foreign oil?

Inflation	Shall we permit unemployment to increase, in order to control inflation?
SALT	Can we insure Russian compliance with the strategic arms limitation treaties (SALT)?

These are all clearly public policy problems, because they are concerned with how society lives. They all also deal with extremely complex scientific issues. Some involve the "hard" sciences, such as nuclear physics and biochemistry. Others involve the "soft" sciences, such as economics and sociology. People seeking the best possible answers, quite naturally turn first to scientists for help. When they do, they often find the experts just as bitterly arrayed against each other as lay persons.

Sometimes two experts will each decry the disaster that will "surely" befall society if it follows the advice of the other. Decision makers in government and the judiciary must then struggle to balance seemingly irreconcilable positions and viewpoints on complex matters about which they have no scientific competence. Increased polarization of the lay population has been a serious consequence.

In a democratic society, it is important that the major part of the interested public be satisfied with any public policy decision. The inevitable result of our present inability to find satisfactory answers to socioscientific problems is that much of the public is confused and uncertain. Some people have become disenchanted with the dispute-resolution process by which their interests are being determined, including the conduct of the participants in that process. We see a growing willingness to solve these difficult issues with simplistic solutions—restrictive central government control, or limitations on scientific research and technological development. Such control or prohibition may be the most serious danger of all. Both threaten the freedom which is the foundation upon which our democracy is built.

Now is not a time for alarmism. We are not far down the road to Armageddon. Science and technology have surged forward since World War II. They have given us the potential to do wonderful things: eliminate poverty, provide a fine education for all, improve health and quality of life, perhaps even find world peace by removing the inequities which are so often the cause of war. To help realize this potential, however, one thing we must do is step back and examine how we, as a society, are dealing with science and technology. This will show that we have not yet developed the social tools needed to resolve the complex modern (often scientifically based) disputes with which we are faced.

Some people consider that dispute resolution involves only litigation. Far from it. The American dispute-resolution process provides for many different ways of resolving disputes. They include discussion, negotiation, and compromise—the ways in which people solve most problems. They include mediation, conciliation, arbitration, and other forms of voluntary accommodation. There may be governmental action on all levels. There are many variations of each of these. All are cast within an adversarial framework, which has a history reaching beyond the founding of our nation into centuries of earlier English roots. They have served us well. Only once in our history has civil war threatened our nation. In the past, our citizens have been quite willing to live comfortably with the decisions which have resulted from these traditional dispute-resolution mechanisms.

Times have changed, however. Many of the critically important modern problems which our society must today resolve—what I have called "socioscientific disputes"—are different in degree, and sometimes in kind, from those which our existing dispute-resolution mechanisms were designed to handle. We are about to enter the twenty-first century, but our scientific dispute resolution continues geared to the adversarial extremes of a much earlier day. With understanding and commitment, we can develop new and enlightened approaches to facilitate the use of science to resolve societal problems. We can then enjoy more of the great benefits which modern science and technology offer.

Society must catch up with science. This book suggests some ways to help achieve that objective.

The Socioscientific Dispute

The socioscientific dispute has three critical distinguishing characteristics. First, there is always a deep and abiding public interest in its resolution. Second, the information and understanding required in order to come to a rational judgment are extraordinarily complex and difficult to evaluate. Third, a sound final judgment requires the fine tuning and balancing of a number of "quality-of-life" value concerns, about which different people may have widely varying attitudes and feelings.

None of these features alone is sufficient to constitute the true socioscientific dispute, as I define it. Many disputes involve one and sometimes two of them. When all three are present, however, they seem to synergize. It is these kinds of problems which existing dispute-resolution mechanisms usually cannot resolve on a basis which is acceptable to large segments of the public. These are the true socioscientific disputes of this study.

The remaining sections of this chapter will discuss and analyze each of these three characteristics of the socioscientific dispute.

The Public Interest

One major characteristic of the modern socioscientific dispute is suggested by its very first term, *socio*. There is high public interest in the outcome of conflicts involving the impact of science and technology on how we live. Even people who are not members of any "activist" group—nor directly affected by the dispute, nor ordinarily deeply concerned with public issues—can become vituperative concerning the issues.

One definition is important at the outset. I will frequently refer to the "public" and the "public interest." I do not mean to suggest by these terms that *all* persons are interested or involved in the issues under discussion. These are simply shorthand ways to distinguish the general population from members of special interest groups. Such groups are found in the consumer and environmental movements, in industry, in government or politics, and elsewhere. It is when these latter groups are able to influence a sufficiently large segment of the "public," that the pressures which are always present in our society result in action. The philosophy of "the public be damned" is no longer acceptable on any basis.

At approximately 4:00 A.M. on March 28, 1979, the Three Mile Island Unit 2 nuclear power plant located about ten miles southeast of Harrisburg, Pennsylvania, experienced a loss of "feedwater," which led to a "turbine trip" and later a "reactor trip." At 7:00 A.M., the State of Pennsylvania was notified. At 7:09 A.M., the regional office of the Nuclear Regulatory Commission was alerted. Pennsylvania Governor Thornburgh advised that preschool children and pregnant women living within five miles of the plant should be evacuated. For a brief period, he urged everyone within ten miles to stay indoors, with windows shut.[2]

The Three Mile Island "abnormal occurrence" immediately became front-page news around the world. Would the fictional meltdown depicted in *The China Syndrome* become a reality? Would the "hydrogen bubble" formed within the reactor explode, blanketing Pennsylvania with dangerous radiation? Which way would the winds be blowing, and where would radiation travel? How hot was it inside the reactor vessel? When might "cold shutdown" be achieved? How serious was the damage to the nuclear fuel rods and to the plant itself? Why hadn't anyone warned that all this might happen? What and who were responsible? Would it happen again? Could nuclear power survive the event?[3]

The incident riveted the interest of the public and generated personal anxiety. Even people who had not been seriously concerned with the Vietnam War or Watts, watched and waited for the news. My own private "public interest" index is based on my youngest son's "one-liner" humor. He came home from college with the following examples of the college community's reactions:

Q. What's the weather forecast for Pennsylvania?
A. 3,000°, and *very* bright.

Q. What's the five-day weather forecast for Harrisburg?
A. Two days.

Humor usually contains at least a grain of truth. Nuclear disaster *is* a serious public concern.

On March 9, 1977, the Food and Drug Administration (FDA) announced its intention to ban saccharin from the U.S. food supply. Once again, the public response was quick and dramatic. Although the FDA cited laboratory experiments which indicated that saccharin must be regarded as a cause of cancer in humans, the public was unconvinced.

> The citizenry, which is often fonder of its frivolities than its necessities, regarded this announcement as the point where the U.S. government finally stepped over the bounds into tyranny. Polls revealed that overwhelming majorities—whether saccharin-users or not—were opposed to such a ban. Adding to the uproar came protests from some professional groups—dentists, dieticians, diabetes specialists—who insisted that the conjectural harm in a substance sanctioned by seventy years of use was probably outweighed by its benefits in curbing tooth decay, obesity, and blood-sugar problems.[4]

One congressman reported that he had "received more spontaneous mail on saccharin than on any other issue in the 18 terms I have served in the House."[5] Others were quoted as saying that "the resulting deluge of mail was second only to the torrent that followed the President's 'Saturday night massacre' at the height of the Watergate turmoil."[6] HEW Secretary Califano "told a Cabinet meeting he was receiving 600 to 800 letters a day, almost all protesting the ban."[7] The result was a congressionally enacted moratorium on the ban, pending a special National Academy of Sciences (NAS) study. When that moratorium expired on May 23, 1979, with no scientific answer that seemed to satisfy the public at large, a further two-year moratorium was enacted by the House, and informally complied with by FDA, pending Senate and presidential action.[8] The existing food and drug statutory framework, and the historical pattern of FDA enforcement, simply had to give way to the overriding public demand.

The Impact of Science on Society

The Three Mile Island and saccharin episodes illustrate the great current public interest in the impact of science and technology on society. Such interest is a relatively new phenomenon. At an earlier time, few persons cared enough to insist on change. Now, more people understand that they may be directly and personally affected. In addition, they have time to reflect on, are interested in, and are able to grasp, at least the general import of what is happening. They appreciate that the impact of technology is now far more rapid, that there is little time to adjust, and little chance for escape. All of this helps contribute to a growing dissatisfaction with the ways in which such disputes are being handled.

Certainly, scientific investigation and discovery, and the technological applications that result,* have had their impacts on society, since the days of the first prehistory discoveries. The discovery of how to make and handle fire, and the invention of the wheel, obviously altered the course of human development. Later developments such as the printing press, the cotton gin, electricity, the automobile, and radio and television, to name but a few, changed the face of this planet.

A century back, only a limited number of people could appreciate this fact. Today, society has matured in very significant ways. We have more leisure, more wealth, and more education than ever before. More people have the time, the capability, and the inclination to deal with quality-of-life issues. The consequent enhanced level of public awareness has contributed to the development of present-day "environmentalism" and "consumerism."

Equally important, scientific and technological developments of the past effected change much more gradually than they do now. It took decades or generations for the printing press or the automobile to profoundly affect the general population. As a consequence, society had time to develop new institutions and processes to adjust to the resulting changes. Those few people who understood the ultimate probable significance of a discovery could also properly expect that there would be time to search for, and to find, satisfactory solutions to the problems envisaged. Thus, with the exception of a few isolated and special instances (such as the destruction of laborsaving machinery by the Luddites in the early nineteenth century), public interest in such matters was minimal.

In contrast, modern technology has narrowed the periods between discovery, initial application, and broad public impact. Many people appreciate that there is often far too little time for existing institutions to react and accommodate. The incredible development of computer technology provides a compelling example.

In a speech delivered at the seventy-fifth anniversary of the Scripps Institution of Oceanography at La Jolla, California, in 1978, National Science Foundation (NSF) Director Richard Atkinson described the impact of the computer as follows:

* "Science" and "technology" are of course quite different terms. When it is not misleading, however, I will sometimes for convenience discuss them together as "science" or as "technology."

> To summarize, I believe we are in the early stages of developments in the science and technology of information processing that will truly revolutionize our society. Advances are occurring at such a fast pace that recent experience is not always a good guide to the future. In the past 30 years, computer computations have gone from a few instructions per second at a cost of several dollars to millions of instructions per second at a cost of less than 1 cent. But such dramatic indicators of progress do not measure the full impact of what is taking place or what is likely to occur in the next 30 years. There can be little doubt that these changes will alter the way people live and earn a living, and the way they perceive themselves and relate to one another. Those changes will have more impact than any of us can foresee today.[9]

Added to this collapsed time frame is an awareness that we may lack the opportunity we once had to at least postpone, and perhaps even avoid, the consequences of technological change. In years past, one could move from Europe to America, from the industrialized East to the western frontiers, or from city to suburb. Perhaps with space travel these opportunities will one day return but, until then, we no longer have such choices open to us. We either do or do not have nuclear power, or saccharin, or genetic research. We cannot wait until all the evidence is in; no decision is still a decision. We must live in the environment we create. There is no effective escape left.

The inevitable result of all this is that more and more persons are seriously concerned with the outcomes of disputes over socioscientific issues. Not all want to, or can, participate, of course. But the effect of this public interest is certainly great. Large numbers of people become directly and even intimately involved; others, although willing to accept a secondary role, still insist on knowing about what is going on. They demand assurances that the ultimate decisions reflect what they regard to be the "public interest." These demands must be satisfied.

Complexity of the Issues

A second obvious identifying characteristic of the socioscientific dispute, is suggested by its middle term, *scientific*. The involvement of scientific issues almost inevitably results in great complexity, at least to lay persons. Indeed, the dispute will often concern not just one, but several different esoteric scientific specialties. Moreover, it usually involves the difficult task of technology assessment. Sometimes huge organizations must be investigated to determine their "intent"—a legal fiction. Sometimes there must be a determination of whether or not there was a conspiracy, another legal perplexity. Even conscientious decision makers have great trouble arriving at judgments on such matters. Most laypeople certainly cannot be expected to understand the merits of what is being said and done. Where the decision affects their interest, or their perception of the national interest, *they will be satisfied with a decision only if they are satisfied with the process by which that decision has been reached.*

The typical modern socioscientific dispute often centers around a large number of highly specialized disciplines and subdisciplines. I have been involved for nine years in one nuclear plant licensing proceeding. Sciences ranging literally from archeology to zoology have been involved. During one session before a three-person licensing board, an ichthyologist, a biochemist, a nuclear physicist, an economist, and a horticulturist all testified within a very few days. Each employed the esoteric words of the Ph.D. Each expressed opinions and conclusions which only someone similarly skilled could adequately evaluate. The two nuclear physicists and the law professor who constituted the board had as much difficulty understanding testimony outside their own specialties as did lay observers.

As twentieth-century civilization and science have progressed, we have come to rely upon an increasing number of such esoteric specialties. There is no longer any Leonardo da Vinci "Renaissance Man" who understands all science—if there ever was. Indeed, it is my experience that a particle physicist, for example, can have a far more difficult time understanding the thinking of a psychologist than does an educated nonscientist. This is because of the specialized nature of the particle physicist's training. Everyone is a lay person outside of his own field.

Technology Assessment

This requirement for scientific expertise, however, is but one of the contributors to the complexity of the modern socioscientific dispute.

Another is the predictive nature of the concerns involved—what we generally call the process of "technology assessment."

Until the nuclear test explosion of a plutonium device on July 16, 1945 near Alamogordo, New Mexico, no one could be absolutely certain that the outcome of the Manhattan Project would not be a chain reaction that would destroy the world. Even then, some doubts persisted regarding the physical consequences of exploding plutonium and uranium bombs. Until the nuclear power plant's emergency core cooling system (ECCS) was tested in actual operation, no one could be sure that it would contribute to "defense in depth," as planned. At some points during development, there are similar uncertainties with regard to substantially all other new products and ideas, from new drugs and pesticides to economic theory and school busing. We are not dealing in these cases with what has been—history—as the issue. Instead, the problem is over what will be, or what would have been. Frequently the conditions are uncertain, the environment hypothetical, and the time undefined. We may be able to simulate the nuclear episode in our computers. Until a catastrophe actually occurs, however, as Three Mile Island demonstrated, we simply cannot know absolutely whether our simulations are valid representations of reality. In the "soft" science context,* we may develop models of what competitive pricing would have been had someone not engaged in unlawful monopolistic practices. Here we can never know whether we are correct. The conditions are strictly hypothetical. The precise events can not be created.

Uncertainty, real or speculative, is a hallmark of these matters. Thalidomide, when first introduced in the late 1950s, was so safe and effective a sleep-inducing drug, that it was sold in Europe without prescription. Soon it turned out to be a dangerous teratogen, which caused the malformation of about 8,000 children before it was recalled. Today, fifteen years later, it has become a "wonder cure" for 12 million victims of leprosy throughout the world. In 90 percent

* The "soft" or "social" sciences, such as economics and sociology, can be just as complex as the "hard" or "natural" sciences. They can be far more uncertain. They can pose more serious problems, because lay people may *think* they understand, when they do not. A lay person will rarely dare to speak about nuclear physics; he will often purport to speak with authority about economics. Some law professors argue that the "socioscientific" dispute should not be defined to include the soft sciences, and they may be right for some purposes. But for the analyses and recommendations of this book, the soft and hard sciences are not different. I include them in my definition of the socioscientific dispute, and they are included in my definition of "science" throughout this book.

of 4,552 reported cases, symptoms of the crippling disease showed improvement in two days, and total remission was achieved in two weeks. Most lepers who once required hospitalization can now be treated as outpatients.[10] From 1930 to 1950, therapeutic irradiation was widely accepted as a medically safe and effective treatment to shrink tonsils, adenoids, and the thymus gland. In the late 1970s, it was found that such treatment may be a cause of cancer. Many of its recipients began developing thyroid cancers and benign nodules on their thyroid glands, as much as 35 years later.[11]

Advocates of any view can project and argue with persuasive effectiveness, and may change positions as conditions mandate. In New York City, the federal Environmental Protection Agency (EPA) and other toll advocates have contended that the imposition of $1 round-trip tolls on the East River and Harlem River bridges, would reduce air pollution by discouraging traffic into the city. The New York City Environmental Protection Administrator contended, in opposition, that "toll booths at the bridges would increase the pollution rather than decrease it because vehicles would have to slow down, wait and then accelerate away from the booths."[12] The *New York Times* for years opposed any effort by Consolidated Edison (Con Ed), the local utility, to burn high-sulfur fuel or coal. In mid-1979, however, it modified its position to approve a one-year test of dirtier fuel.[13] Times and circumstances had changed sufficiently, in its opinion, to justify a different approach.*

The relatively new and developing science of toxicology seeks to ascertain the effects of substances (and conditions) on living organisms. Law is concerned with a sometimes simplistic concept of "proximate cause," by which liability is assessed. In contrast, toxicology evaluates each of the numerous factors which usually play a part in a particular end result. Even though each may constitute a proximate cause in law, not all are by any means regarded as directly

* In an editorial as the decade turned, the *New York Times* commented on the U.S. government's switch from 8-x-10 ½ inch to 8 ½-x-11 inch paper, the size used by the rest of the country. The narrow paper size was ordered by commerce secretary Herbert Hoover in 1921 as an economy measure; shaving margins meant using less paper. Over a half century later, a GSA study committee concluded that the larger size paper increases the typing text area by as much as 20 percent, while increasing sheet size by only 11 percent. Thus more documents could be prepared with fewer pages, at an anticipated overall 5 percent saving in the number of pages used. One savings calculation was $7.5 million a year in clerical labor, plus another $3.5 million in storage costs because thinner reports would take less space. AID, however, took issue with GSA, reporting that the new size paper is too large to fit the stationary drawers of many standard GSA desks. So AID will have to exchange desks, provide desk-top stationary racks, and resort to other costly devices—which far exceed the calculated savings! Editorial, "More is Less, More or Less," *NYT*, January 1, 1980.

causative in the toxicological sense. Where law and toxicology interface, as they so often do in socioscientific dispute resolution, the result can be even greater complexity in the assessment process.

The toxicologist trying to determine whether, and under what circumstances, a proposed new chemical compound or physical exposure will "cause" harm to humans may compare it chemically with other substances whose effects are known. He may perform *in vitro* tests. Neither of these will usually provide an acceptable answer. Human experimentation is ordinarily not permitted by present morality. Biological tests are often inadequate. As a consequence, virtually all toxicologists today agree that animal experimentation must frequently be conducted.

The toxicologist must decide on the number of animals to be tested, the sex and species to be used, the duration of the tests, the levels of exposure, the routes of exposure, the parameters to be investigated and when, what control animals are required, and an almost infinite number of other matters. His decision on each of these will turn on a great variety of considerations, including the probable nature of the human exposure, the anticipated benefits of the compound, and the character of the potential risk. The ensuing tests can take years, require the skilled expertise of many different specialists, and cost hundreds of thousands or even millions of dollars.

When all is said and done, however, everyone knows that animals are different from humans, some more so than others. Tests for two generations are at best only suggestive of what can be expected to happen in a third, or subsequent, generation. What is observed in one hundred or one thousand animals can not establish beyond doubt what will happen in a population of one million or one billion. What is unseen may mean only that the analytic capability is not sufficient to the assignment. The result is that toxicological predictions are "guesstimates," expressed in statistical confidence levels, and including some measure or degree of risk which even experts can debate. In toxicology, as elsewhere, absolute certainty of harm, or freedom from harm, rarely exists. Where experts include "quality-of-life" judgments in their opinions by asserting that something is "acceptably safe," or "unreasonably dangerous," debates between them can generate the heat and venom which so often confuse and dismay the lay observer.

Organizational Intent

"Organizational intent" is the intention, or motivation, or purpose, of an organization, as distinguished from an individual. This is another important contributor to the complexity which characterizes the resolution of socioscientific disputes.

Usually, a large corporation or a government, or both, will participate as parties to these disputes. The motives of one or more of them will be relevant to the outcome of the controversy. An organization itself, of course, has no mind as such. We must ascertain the conduct of the people in the organization to ascertain its intent. Hundreds or even thousands of persons may be involved in making and implementing decisions.

Nonlawyers who are not participating directly in socioscientific disputes often do not understand the importance of "intent" to the decision process. This is especially true of scientists, to whom intent is ordinarily irrelevant (except where intent is itself the scientific issue, such as in psychiatry.) Criminal guilt and innocence, and civil liability in damages, in fact frequently turn upon this issue. In a civil rights case, the question may be whether the organization's intention in refusing to promote a member of a minority group was to discriminate against him (a "bad" and therefore an unlawful intention) or to reward someone else for effectiveness (a "good" and therefore a lawful intention.) In an antitrust case, the question will be whether a special trade practice was adopted by a company for the purpose of competing against an aggressive competitor (a "good" purpose, and therefore "lawful"), or in order to monopolize the industry ("bad" and "unlawful"). In a government tax prosecution, the question is whether a tax agent served a civil administrative subpoena as part of a valid civil tax investigation ("good" and "lawful"), or in order to obtain evidence in support of a planned criminal prosecution, and therefore in violation of the taxpayer's constitutional rights ("bad" and "unlawful").

In political, commercial, and social disputes, as in the law, we often look at the giant organization as if it were a person—and assume that it has the same kind of mental process. This anthropomorphism is a fiction, and "organizational intent" is a fiction as well. Ordinarily, such intent is in fact created retroactively by examining all the bits and pieces of evidence which can be found, and then deciding what must have been the organization's thinking, or purpose, in doing what it did.

Consider, for example, a government or private antitrust litigation against a large corporate defendant. This kind of case is one of the most difficult of all socioscientific disputes to resolve using present dispute-resolution mechanisms. Frequently, one of the important issues will be whether or not the corporate defendant had an "intent" to restrain trade or to monopolize.

Suppose that years before the alleged misconduct, the corporation's board of directors had adopted a clear and positive resolution, or even a bylaw confirmed by the stockholders, that there was to be absolutely no anticompetitive intent or conduct at any time. Suppose that it communicated this to all its personnel, that it kept both the instruction and its communication current, and all company personnel carefully cleared whatever they did with the legal staff. I am confident that the defendant's own trial attorneys would agree that this history, although relevant, would certainly not be determinative of the "intent" issue. Intent would still be decided by examining all the acts and statements of company personnel over the years, and, then only, deciding what must have been the intention that motivated it to act. Otherwise, one would usually find such an instruction in the records of any well-run corporation, and this is not the case.

A Supreme Court decision at the end of its 1978 term is most revealing as to the fictitious nature of this organizational intent issue. An agent of the Internal Revenue Service, clearly seeking to develop a criminal prosecution of a taxpayer, served a civil subpoena to obtain the taxpayer's records. The subpoena was attacked as unlawful, on the ground that it had been improperly served in order to obtain evidence for a criminal case. The Court agreed that the subpoena would have been unlawful had it been served for such a purpose. However, it carefully distinguished the intention of the government, as an organization, from the intention of the government's agent—even though the agent was the only human being who was directly involved, and had the only real life "intent." The agent was only a single person, the Court said, and there had not yet been any organizational government decision to prosecute (nor perhaps could there have been without the evidence being sought by the subpoena.) Therefore the government's intention was not improper, however bad the agent's might have been, and the subpoena was valid.[14]

To the taxpayer, this may have seemed to be a "Catch 22." He had a right to prevent compliance with the subpoena if it was to be used for a criminal prosecution, but it was not to be used for the criminal prosecution until after he had obeyed it. Therefore, he had no right to prevent compliance with the subpoena.

This need to determine organizational intent is critical to the decisions in many major socioscientific disputes.[15] It is often the primary reason why their resolutions require the examination of millions of pieces of paper and other records, large numbers of witnesses, great lengths of time, and enormous expense. The U.S. government's antitrust case against IBM Corporation provides the classic example. It had its tenth anniversary on January 17, 1979, at which point the government had reviewed 60 million pages of IBM documents and IBM had inspected 26 million pages of government documents. By its fourth anniversary of actual court trial—on May 19, 1979—in addition to 1,309 witness depositions, there were some 90,000 pages of trial transcript, with the end nowhere, even remotely, in sight. Then, on June 25, 1979, the judge directed IBM to produce what it estimated to be an additional five billion documents, which IBM claimed would involve 62,000 man-years of work and $1 billion production cost. This, among other things, provoked a 2,000-page, nine-volume application by IBM for the substitution of a different judge to preside over the balance of the case.[16]

The cost in time which results from procedures of this kind can mean that the final decision will come only after the problem no longer exists. The expense and other burdens can be so great that only the powerful and wealthy can afford to prosecute or defend their rights. The societal problems caused by this single aspect of the dispute-resolution process sometimes seem greater than the benefits of coming to a decision about the conflict.

Conspiracy

The additional issue of conspiracy in many socioscientific disputes adds still another element of complexity to their resolution. "Conspiracy" implies joint action by two or more persons or organizations. In many cases, society has decided that such joint action is improper or unlawful, even though the same action, done independently, is perfectly proper. A common, and very simple, example is price-fixing. Any competitor must, of course, decide on the prices at which it will market its products, and usually this is perfectly lawful. When two competitors agree on the prices they will charge, however, they may not only be acting unlawfully, but can be guilty of criminal misconduct for which those involved will be fined and imprisoned. In the mid-1950s, the major automobile manufacturers agreed to cross-license certain antipollution developments to each other. Had each acted individually and alone, there might have been no objection. A decade later, however, the agreement to act jointly and together led to antitrust prosecution. The government's objection was that the agreement to share developments inhibited research and development, because it meant that no one could gain competitive advantage by virtue of a new discovery that it could use exclusively for itself.[17]

As with "organizational intent," conspiracy itself is very often a fiction, also determined retroactively by an examination of everything that has been said and done. It can be based upon what "must have been," rather than upon anything that was actually spoken, written, or spelled out by the parties involved. Like the hypothetical corporate "no anticompetitive intent or conduct" resolution, sometimes the parties to the conspiracy will themselves fail to appreciate that they have in fact "conspired," and will in all good faith protest their innocence. Even in those cases where the conspiracy is overt and real, by its very nature it will still be secretive and clandestine. Thus, with rare exceptions, it can only be established by indirect and circumstantial evidence, in much the same fashion that organizational intent is shown. All this adds to the complexity of the issues.

With all of these characteristics of complexity—the involvement of multiple sciences, technology assessment, organizational intent, and conspiracy—it is little wonder that even skilled scientists have great difficulty in understanding all of the issues involved in modern socioscientific disputes, and that it is so expensive and otherwise burdensome as to preclude the effective participation of many who want "in." Often it takes so long to understand and decide, that at the time of final resolution (if it ever comes), the problem is no longer the same as the one which has been analyzed. This is largely the fact in the IBM litigation. The computer industry of the 1980s is vastly different from what it was when the case was begun over a decade ago. The technology includes very large-scale integration, mini- and microcomputers, distributed data-processing and data-base operations. There have been radical changes in production and marketing practices. Many old competitors are no longer around, and some new ones are thriving. Yet one ultimate issue in the case continues to be whether and how that industry is to be restructured in light of IBM's conduct in 1969 and before. When he left office in mid-1979, Attorney General Griffin B. Bell said that the duration of the IBM case suggests that there may be "something wrong" with the court system.[18] Federal appeals court judge William H. Mulligan, during argument, stated that the trial "is lasting longer than World War II" and "has become a Frankenstein monster."[19]

Decisions which are not easy for even the most skilled decision maker are obviously, and necessarily, far beyond the ken of the lay person and the public. The public cannot understand the merits of what is going on where dispute resolution is concerned. Thus, any doubts the public has as to whether its interests are being properly considered and served can be satisfied only if it has confidence in the integrity of the dispute-resolution process being employed.

Risk/Benefit Analysis

The third and most difficult identifying characteristic of the modern socioscientific dispute is the need to make "quality-of-life" decisions. This calls for the application of a relatively new and developing methodology, commonly called "*cost*/benefit analysis." I prefer to use the term "*risk*/benefit analysis," because it avoids an unfortunate implication that health and lives are being equated with money.

Risk/benefit analysis means that the benefits of a given course of action are compared with its "disbenefits" (by which is meant *all* risks, and all costs, not just money.) Once this is completed, a decision is made based on which appears to be the greater. This latter process is sometimes called "solving the risk/benefit equation."

Risk/benefit analysis can be extraordinarily difficult to perform satisfactorily. This is because it recognizes that the "quality-of-life" issues involved in socioscientific decision-making are really to be decided by personal and societal value judgments as to how we wish to live. It asks questions such as whether we are willing to accept greater air pollution in order to enjoy the benefits of greater automobile travel. It tells us that answers in these matters come only from comparing what we want with what we do not want, following careful analysis of all the probable consequences of the given course of action.

The Historical Setting

The risk/benefit concept is a comparatively new one. Risk/benefit analysis—the related methodology—is a new process. It will be helpful to examine the concept in its historical perspective to identify some of its important special features. This will clarify differences from earlier "absolutist" approaches to the resolution of science-impact issues. Most important, it will help explain why risk/benefit also calls for the development of new approaches to dispute resolution.

World War II marked a turning point in the public's perception of the impact of science and technology on society.[20] Before, relatively few persons appreciated the enormous social significance of scientific developments. Too much time elapsed between cause and effect for people—even those with time, education, and interest—to pay much attention to such matters as the probable effects of Albert Einstein's discussion of the equivalence of matter and energy. His formula $e = mc^2$, which led to the development of atomic power, preceded the atom bomb by three decades.

World War II began the process which changed this. During the war, the industrial nations poured resources into research and development without effective limit, other than that imposed by human and material capacity. Enormous advances occurred in physics, chemistry, biology, astronomy, mathematics, and the other sciences, leading to such applications as nuclear power, computers, space travel, laser technology, genetic manipulation, and many others. Although there are still substantial time gaps today between discovery and broad public utilization, one important result of these advances is that the periods are far shorter. Sometimes, a new physical or chemical application will call for societal adjustments which seem to follow hard on the heels of the relevant discovery. The most obvious examples are in the computer field, where developments such as the "chip" and inexpensive mass memories have already changed the ways in which we do business and set us on the path to electronic fund transfers and what many still choose to call the "checkless/cashless society." The consequence of all this has been a growing public awareness of—and appreciation for—the impact of science on society, and an awareness of the need to develop suitable societal mechanisms in response.

It is usually dangerous to treat recent historical developments from limited or special perspectives. But if it is recognized that there were many other trends, and many other causes and effects, it will be useful to divide our postwar years into three periods in terms of societal response to science impact.

The first and second periods were somewhat simplistic and absolutist, swinging first to the far political right and then back to the not-quite-so-far political left, in typical societal pendulum fashion. The third period is the "risk/benefit" period of the present. It represents a centering of the pendulum, and, in my opinion, a far more sensible, reasonable, and responsible approach to the problem.

The First Period: Optimism

On August 6, 1945, the world was shocked by the first dramatic, publicly effective demonstration of the effect of science on society—the dropping of the atom bomb. Initial reactions were colored by the realization that the war was over, and that the soldiers were coming home. In a short time, however, the far deeper significance of Hiroshima began to sink in—hundreds of thousands of humans could also be killed in an instant, in New York and Aberdeen as well as in far distant places. Whatever euphoria might have previously existed was soon replaced by the sheer horror of what had happened, and the terror at what might happen, anywhere, anytime.

There was general appreciation that international society had not yet developed any adequate controls to insure that atomic power would be employed only for the good of society. The United Nations was seen as a brave experiment, but in the tradition of the League of Nations and the World Court, with no greater assurances for success. No one could yet rely on a community of nations with effective police power. One understandable response was continued preparation: offensive by way of the arms race, which continues even today; and defensive by way of civil defense. Although little of the latter remains today, civil defense seemed almost to proliferate after the war. When my wife and I moved from our house in 1977, our basement still had disaster provisions in it, including two fifty-gallon jugs of drinking water which had been stored there in the 1950s. My NYU classrooms still had the familiar air-raid warning signs on the walls, telling where to go and what to do in the event of an attack.

None of this furnished a completely satisfactory answer, however. Whatever else could be done to prevent atomic holocaust, should be done—but what? The problem was worldwide, but the response in the United States was different from elsewhere. We were, after all, the only atomic "have" nation; everyone else was a "have-not" country.

Perhaps today the American public appreciates that the matters of world security and nuclear control are complex problems, which may never be solved to the full satisfaction of everyone. But in the first postwar period, many were ready to believe that there was an easy answer—secrecy. The analysis was really quite simple: We alone know how to make the atom bomb. No one else does. We will certainly never use the bomb improperly. If we can only protect our secret, everything will be fine.

The decade that followed explosion of the bomb saw the Smith Act and the Attorney General's compulsory registration list for "subversive" organizations, "red" witch hunts, McCarthyism, and much else which took place in the name of "loyalty," fighting communism, and avoiding atomic war. It may seem strange to call this a period of "optimism," but that is what it was in terms of the public's perception of the science-impact problem. There was widespread belief that society could control science. Relatively few persons appreciated that we have no permanent monopoly on scientific information. Even fewer understood that other different science-impact issues were soon to emerge into the forefront of public consciousness, and demand public attention.

The Second Period: Pessimism

There is no clear or fixed dividing line between the first two postwar science-impact periods. The first, of optimistic belief in the ability of society to control science, began to fade away toward the end of the 1950s. By then, time had dulled the fears of atomic disaster, and the reaction to McCarthyism had set in. Food and drug, water pollution and other legislation had been on the books for decades, limiting or prohibiting certain types of activity.[21] These provided a setting for the next science-impact period of "pessimistic" belief that society could not manage or control science satisfactorily, and that therefore additional and far more pervasive limitations and prohibitions were the only viable means of survival. Perhaps the greatest single impetus to this developing feeling of pessimism resulted from the publication in 1962 of Rachel Carson's immensely popular *Silent Spring*. Thereafter it was fed and nurtured by the warnings of Ralph Nader and a growing number of other environmental and consumer leaders and organizations. All told of the dangers of pollution, depletion, overpopulation, carcinogenicity, and a virtually unlimited number of other potentially negative consequences of our so highly industrialized society.

Once again, the answers furnished were somewhat simplistic and absolutist, although this time at the not-quite-so-far political left end of the spectrum. More and more people began to call for a total ban on nuclear power, a prohibition on the use of most agricultural chemicals, and specific and well-defined limits on industrial growth and population. This approach is reflected in the "Delaney Clause," a Food and Drug Act amendment interpreted to prohibit all food additives causing cancer in laboratory animals, no matter how small the cancer risk to humans or how great the benefit.[22] It is also found in the original Endangered Species Act (ESA),[23] which could prohibit any project (such as the Tellico Dam) which jeopardized an endangered species of life (even so insignificant a one as the snail darter minnow). It made no difference how important or advanced the project in construction, or how trivial the species involved.

Many of the same "activists" who were disenchanted with the ability of society to control science were also prominent in the struggles against the Vietnam War, the student riots, and other bitterness which characterized the period. The product of it all was clear and growing public disenchantment with science and technology and those supporting them, including government and corporate society. There was a conviction that change was necessary to avoid disaster. That change was essentially to halt science and technological development in their tracks—an assignment which is no more practical or possible than it was to keep such matters secret.

The Present Period: Realism

In 1969, Congress passed the National Environmental Policy Act (NEPA),[24] almost by inadvertence and certainly with no serious discussion of what it was doing. It contained a provision mandating an "environmental impact statement" for any major federal action affecting the environment. That statement was required to assess all the risks and all the benefits of the proposed action, before the action could be taken.

Whatever might have been the Congressional intent, by its language NEPA seemed clearly to call for a "risk/benefit" approach to the solving of environmental problems. This was in obvious sharp contrast to the absolutist approaches of the two earlier postwar science-impact periods. Soon, following a number of court cases in the early 1970s, the real significance of NEPA began to sink in—first among government and industry representatives, some scientists and a few others, and then a much larger segment of the general public. People who were not members of any directly affected organizations or groups, began to talk in terms of "balancing the risk/benefit equation," and the need to make decisions maximizing society's interests after evaluating all the pluses and all the minuses. Acceptance did not come easy, however. Even today, there are many in all camps who object to the risk/benefit approach as "weak," or "temporizing."[25] Despite this, the balancing approach is being reflected in an increasing volume of legislation. This includes, most recently, the very important Toxic Substances Control Act (TSCA),[26] which regulates a broad spectrum of chemicals and the chemical industry, and the amended Endangered Species Act,[27] which relaxes the earlier absolute prohibitions. It also includes the special statutes regulating cigarette sales and temporarily forbidding any ban on saccharin as a carcinogen ("Warn the public properly, by requiring clear notices on these products as to their potential health hazards, but don't prohibit people from taking risks they want to take.") In mid-1978, the Office of Manpower and Budget went so far as to introduce the concept into its guidelines for preventing fraud in federal assistance programs. It directed action "balancing the government's need to maintain the integrity of federal programs with the individual's right to personal privacy."[28] Also, Massachusetts Senator Kennedy's proposed antitrust standard for federal agencies prohibits any anticompetitive agency action unless it is first demonstrated that "the anticompetitive effects of such action are clearly outweighed by significant and reasonably certain benefits to the general public."[29]

Recognition and endorsement of this new approach to science-impact problems cuts across virtually all fields of science. Anna J. Harrison, past president of the American Chemical Society, put it this way:

> Sciences are amoral.
>
> The use of science and technology may be judged by society to be moral, amoral, or immoral.
>
> All technological change (all change for that matter) has a positive as well as a negative impact on some segments of society.
>
> There are things science and technology can accomplish, and things that science and technology cannot accomplish (i.e., create a risk-free technology).
>
> Innovation is a slow process based upon technological competence but highly dependent upon social, economic, and political factors.[30]

Microbiologist René Dubos, one of our great science-humanists, has pleaded for balance, to permit the creation of "environments which are ecologically stable, economically profitable, esthetically rewarding, and favorable to the continued growth of civilization."[31] Economist John Kenneth Galbraith has urged understanding that protection of the environment has a cost. Sometimes damage must be accepted in the interest of society overall.[32]

We have a long way to go before we will have satisfactorily solved the serious problems which the advances of modern science and technology pose for society. There may be new and different periods of response. However, we have come to a better understanding of what we must do about socioscientific issues. There are few easy or absolute answers. We must tote up all the risks, and all the benefits; next, we must compare them. We can then decide what is in the overall best interests of society.

Science is neither good nor bad. It is how we use it that counts.

Solving the Risk/Benefit Equation

At least we now know many of the right questions to ask, and that is necessarily the first step. But it is helpful only in setting us on the right track, for it furnishes no answers at all—only further questions, and most difficult and sensitive ones at that. These questions make it clear that we are almost always dealing with human "values." What is a risk or cost, and what is a benefit, are not always simple to determine. The weights given to each vary from person to person, from place to place, and from time to time.

Consider, for example, the matter of how much to clean up an almost totally polluted lake. Eliminating the solid wastes will be relatively simple, by somehow straining out or sedimenting the suspended solid matter. Eliminating additional impurities, however, gets more and more difficult, depending on how far you go. The effort to get the lake 75 percent "pure" might be three times as difficult as that needed to remove the solid waste; 90 percent, ten times; 99 percent, fifty times; 99.99 percent, two hundred times. To remove the last molecule or atom of contamination, might mean treating all the water in the lake on a continuing basis, especially when it rains or smog enters its atmosphere. The effort required would be astronomical, the costs unheard of.

Isn't it obvious that how much you decide to clean up the lake will depend, at least in part, on what you want to do with it? Look at it? Skate on it? Fish in it? Swim in it? Drink it? Use it for automotive battery fluid? Use it, untreated, for critical laboratory experiments?

Isn't it obvious also that what resources you will decide to deploy will depend, at least in part, on what resources you have available, and what you might otherwise be doing with them?

Won't your decision also turn on what alternatives are available to you? If there is another lake nearby which will satisfy your need for beautiful scenery, or to ice-skate, fish or swim, or a sufficiently pure water supply available for drinking, battery fluid, or laboratory experiment, you may decide to leave your lake "as is." When cyclamates were removed from the marketplace as an artificial sweetener, there was relatively little protest from the general public, because the saccharin alternative was readily available. When the threat of removing saccharin was announced, however, the public clamor became much greater, because there no longer was any other artificial-sweetener alternative left.

Finally, if the decision is that of your community, and not yours alone, won't you also need to consider the interests of those along the river into which the lake empties? Also others even more remotely involved or affected, such as persons at the perimeter of the drainage basin along the lake? Also those whose tax contributions will be needed to help defray the effort? Their values and interests may be quite different from yours. Some of them may even be citizens of foreign countries.

Modern socioscientific disputes pose all of these problems, and many, many more. Frequently they call for judgments as to risk, cost, and benefit involving large numbers of people throughout the world, including those in seriously depressed areas, and those with different religions, different economic systems, and different social structures. Many of the effects to be considered will be far down the chain of causation, and much more remote than the immediate health or environmental hazard, or the potentially inflationary or food-production consequences of a proposal such as the banning of a pesticide. What will be the reaction of citizens in other nations? Will they simply manufacture more of a product we ban, to replace what we do not supply? Will an action influence our negotiation of a strategic arms limitation treaty (SALT)? Will it affect world climate and weather years hence? It is difficult enough for one person alone to make such a decision, with all the options and consequences fully laid out. It is far more difficult where all of international society may be involved, including many persons who simply cannot understand what is being discussed.

Legislative Decision-Making

At the outset of this discussion of risk/benefit, I said that of the three special characteristics of the modern socioscientific dispute, risk/benefit poses the most difficult problems of all for our traditional dispute-resolution mechanisms. This is because risk/benefit dispute resolution is inherently, essentially, and unavoidably a "legislative" process. This characteristic is not changed simply because we usually try to handle risk/benefit dispute resolution by a nonrepresentative process such as executive order or lawsuit.

As in the ordinary case being decided by a government official or judge, the risk/benefit arbiter starts with an examination of a mass of conflicting evidence and contentions, and interests—but there the analogy ends. With only limited exceptions, the government official and the judge in the usual case has a specific standard to apply, found in the applicable statute or in the common law. Where that standard of conduct is not sufficiently specific, the action will be struck down as unconstitutional, because it is based on an unlawful delegation or usurpation of legislative power.

The risk/benefit arbiter has no such specific standard. His assignment is simply to decide what he believes to be the best interest of society overall. He examines the "quality-of-life" and "value" judgments of everyone who is or may be involved, and then decides who "wins" and who "loses," or somehow compromises to a third position. If he is given the task of deciding how much to clean up our lake, he will consider what each of us wants or needs; what we are willing to, or should, give up to get it; and what will best serve the community. In the past, in our American system, we have usually reserved this kind of decision-making as the exclusive province of the democratically elected legislator—selected by, and responsible to, the people whose quality of life he is deciding.

I do not suggest that all socioscientific disputes should be decided by legislatures. The legislative process is not competent to handle such an additional assignment. The task is already much too onerous, and will surely become far more so as science, technology, and society progress.

If we are to permit such decisions to be made by other mechanisms, however, we must be sure that those mechanisms are suited to the job. Up to this point, we do not have the necessary assurances. Indeed, we have hardly begun to analyze the problem. There is precious little appreciation even that modern socioscientific dispute resolution is not the same as the dispute resolution of the past. Those persons who have worked in this area have tended to do so along the traditional lines. This means, for example, classification and treatment by the separate substantive rights and issues which are involved, such as antitrust, or environmental, or civil rights concerns; or by the types of dispute-resolution mechanisms which are employed, such as legislative, executive, administrative, judicial, or accommodation. There is no separate academic or commercial discipline treating "socioscientific disputes" as such, irrespective of the kind of dispute or the forum involved.

We need a new and distinct dispute classification, the "socioscientific" dispute, to be described, discussed, analyzed, and debated in a separate discipline. This will permit study of these disputes in terms of characteristics cutting across the standard lines of classification. Such a classification and analysis will not only confirm the serious need for new or revised resolution mechanisms to handle these kinds of conflicts, but will, at least, point to what the general nature of the changes must be. It will permit us to develop new dispute-resolution techniques.

There are many public-interest disputes which are not complex. There are many "complex" disputes which are private ones, and in which all of the important parties are directly and intimately involved; some of these even involve "risk/benefit" assessment. But when a single matter involves the public interest, is complex, and requires risk/benefit analysis, as is the case in the modern socioscientific disputes discussed in this chapter,* these three factors combine with synergistic effect to pose presently insoluble problems for traditional dispute-resolution mechanisms. The public does not care that the rules are carefully and properly followed, which is the primary focus of our traditional adversarial mechanisms, as discussed in the next chapter. The public has great interest in the outcomes of these disputes, which involve important "quality-of-life" problems. It cannot adequately evaluate those results, however, because of the enormous complexity and uncertainty which are always involved. As a result, the public will be satisfied with, and accept, the decisions in these disputes only if it has confidence in the integrity of the process by which those decisions are being reached.

* Most courts, administrative agencies, commentators, and others employing the term "risk (cost)/benefit," apply it almost exclusively to environmental disputes. As suggested at the outset of this chapter and elsewhere, however, I believe that many monopoly and civil rights cases, and some others, also involve the same careful balancing of conflicting societal interests. This is a kind of legislative decision-making. It is just as much risk/benefit assessment as in any environmental dispute. I include these other cases in my discussion of socioscientific disputes in this book.

We turn now to a discussion of dispute-resolution process. The time is similar to that when our colonial ancestors were governed by King George III. They had little confidence then in the methods by which decisions affecting them were made, and protested "taxation without representation." There is little such confidence today. It may not yet have been voiced, but there is growing concern that "quality-of-life" decisions are being made without fairness to the general populace. I fear there is good reason for this. Even if the accusing finger cannot be pointed at anyone specific, such as the British, we may be coming to a period of protesting "socioscientific decision without representation." I hope we can come to solutions in a more peaceful fashion than did our ancestors.

OUR ADVERSARIAL NATION

On June 8, 1979, the press, radio, and television widely reported that the Appellate Division of the New York State Supreme Court had overturned the first-degree murder conviction of Anthony "Tony Pro" Provenzano, described as an "influential Teamsters union figure and reputed Mafia captain."[1] There was no substantive question as to whether Provenzano had done what he was charged with doing. The ground for reversal was that Provenzano had been denied procedural legal rights to which he was entitled. A lower court trial judge had improperly refused to excuse a potential trial juror, who had met the prosecutor at a political club. To get rid of the juror, Provenzano's attorney was forced to use up one of the limited number of "peremptory" juror challenges to which he was entitled by law. As a result, he lost a challenge he might later have used against one of the twelve finally selected jurors who voted unanimously to convict.

The appellate court ruled that Provenzano had been denied his right to a "fair" trial, and that this required reversal. It made no difference at all whether the evidence against Provenzano was sufficient, or even overwhelming; nor whether the twelve jurors who decided were "fair;" nor whether the conviction was clearly "right." The media accepted the decision without negative comment, except for limited complaint that the trial judge should have known better.

On May 25, 1979, just two weeks earlier, American Airlines Flight 191, a DC-10 bound for Los Angeles with 271 passengers aboard, lost one of its three engines on takeoff from Chicago's O'Hare Airport, and crashed, killing all aboard and 4 people on the ground. When the Federal Aviation Administration (FAA) failed to act quickly or firmly enough, the Airline Passengers Association filed suit in federal district court in Washington, D.C., asking that the FAA be ordered to ground all DC-10s until the cause of the accident had been determined. The FAA's attorneys resisted on procedural grounds, just as did Provenzano. At almost the very same moment that Provenzano's objections were being sustained, FAA attorneys were arguing that the district court had no power to require grounding. They too were right; although the lower court issued the injunction, the United States Court of Appeals for the District of Columbia Circuit quickly decided that it had jurisdiction in the case, rather than the district court. The lower court was wrong.

This time, however, the public reaction was not so favorable to the "technicalities" with which the FAA attorneys were defending. *Fortune* reported the lower court injunction proceeding in this way: "By way of counter-argument, the FAA lawyers *merely* insisted that the grounding of airplanes was not the legal province of the district courts. *They did not even try to advance a substantive argument to assure the court of the DC-10's safety*." (italics added)[2]

What is the difference between these two cases? Why in one are we willing to accept, without serious objection, the reversal of a criminal murder conviction on strictly technical grounds and without any apparent concern as to "real" guilt or innocence, yet in the other unwilling to permit even a federal agency to escape improper court jurisdiction over it without assurances that what it is proposing to do is right, on the merits? The answer does not lie in the differences between the parties, private and public, or between a criminal and a civil case. The answer is because we are most reluctant to permit a process which emphasizes form and procedure over result to be employed in a matter where we are so concerned with that result. The DC-10 crash was the worst plane crash in U.S. history. Its causes were complex and uncertain. Much of the traveling public was terrified and demanding immediate corrective action. Automatic, inadequately considered grounding, however, would confuse and upset passenger traffic worldwide, and thereby might unnecessarily and unreasonably create serious commercial and personal problems for would-be travelers.

I may not be happy about it, but I am at least willing to accept Tony Provenzano's acquittal even if I am convinced that he acted as charged. That is the way to assure due process for others, who may be innocent. The immediate and personal effect on me is small, and he can be watched, or imprisoned (as he was) for another crime for which he had been properly convicted. Perhaps the threat he poses can be somehow otherwise nullified by law enforcement. But I am not satisfied to travel, or to let my family travel, in a DC-10, where there is a risk that one of its engines will fall off. Nor am I willing to see DC-10s unnecessarily grounded, at enormous world competitive disadvantage to American airplane production, and at a cost of millions of dollars daily—the niceties of procedure or technicality to the contrary notwithstanding.

The DC-10 dispute is a typical socioscientific dispute. There is public interest in its outcome. It is complex. The benefits of travel must be balanced against the hazards. People are far more concerned with result than they are with procedure. They are not satisfied that our procedures are sufficiently trustworthy for this kind of a case. They do not want to fly in a dangerous plane no matter how fair the procedure was that resulted in the "fly" decision. Their view of the dispute-resolution mechanism is almost the exact opposite of what it is in the traditional dispute, such as in *Provenzano*, where outcome and result are secondary to procedure.

The fatal defect in present socioscientific dispute-resolution methodology is this misplaced emphasis on procedure. It results from blind application of our American "adversary process" or "adversary system" to these new and very different kinds of disputes. Changes are needed if the system is to work satisfactorily.

The adversary process of dispute resolution is the framework within which almost all disputes are resolved in this country, socioscientific and otherwise. That process contemplates that those parties who have serious differences with each other will battle the matter out. Each will employ whatever weapons society allows. Many people believe that the weapons and techniques of the adversary process are used only by private lawyers, or in disputes being decided in the courts and the administrative agencies. That is not the case. Government and "public-interest" attorneys use these methods just as much. Courts and agency tribunals are only the final stages of dispute resolution. The adversary process may be, and is, employed by anyone, anyplace, at any stage. It reaches back to the beginnings of disputes—back to the point where real confrontation first enters the picture, and frequently well before lawyers have been retained.

So long as there is more than one person on an unowned and isolated desert island, disputes between people and their organizations will be inevitable. Husbands and wives, parents and children, buyers and sellers, employers and employees, landlords and tenants, neighboring nations, and all other groups who must deal with each other in one fashion or another, have their differences. These must be resolved if things are to proceed sensibly. Fortunately, they are usually disposed of by discussion, negotiation, compromise, and settlements of all kinds. Were it not so, life would certainly be very different.

In some cases, voluntary resolution does not come easily. Disputes get overheated, and positions become hard and fast. Then the "big" weapons of the adversary process, and its extremes, begin to appear. A huge variety of tactics having little if anything to do with the merits of the underlying dispute are characteristic. These include resorts to technicality, delay, obstruction, "playing it close to the vest," and "forum shopping" for the most favorable judge or court. Sometimes the process seems almost to run wild. One example is the implied charges of homosexuality and antisemitism levied against Ralph Nader during his dispute with General Motors over the safety of the Corvair.[3] Another is the innumerable hearings, appeals, and other battles, which have kept the issue of the Consolidated Edison Company's proposed "pumped storage" power plant at Storm King Mountain still undecided after sixteen years. At Storm King, capital costs alone escalated from an estimated $300 million in 1970, to $1.3 billion in 1979.[4] Anything goes, so long as it is not actually deemed unlawful by some statute, or unethical by some recognized code of conduct.

The adversary process is an integral part of our heritage. Use of the full panoply of adversarial weapons is as common to public interest disputes as it is to commercial, personal, or political fights.* Indeed, the socioscientific dispute is an important one, by definition. It is commonly turned over to attorneys at a much earlier point than the ordinary private dispute. As a result, the attorney's special expertise in exploiting the weapons of the adversary system characterizes these matters almost from their inception.

What is wrong with this? The adversary system has served us well, even with its extremes, for hundred of years. The system has its problems, of course, but so do all other methods of dispute resolution. The public has been quite willing to accept its results in ordinary matters. It will not do the same in complex public-interest affairs, involving quality-of-life decisions, however. This is because the adversary process seeks "justice" and "fairness," not "truth."[5] In socioscientific controversies, the public wants the focus to be on "truth" alone. This distinction is an important one.

"Justice," the object of the adversary process, is defined in terms of the procedures which are employed in the dispute-resolution process, rather than by whatever the particular result may be. This makes it possible to measure "justice" by specific objective tests, rather than by some vague standard, such as what is "right." Webster's *New Collegiate Dictionary* defines "to do justice" as "to treat fairly or adequately," "to show due appreciation for," and "to acquit in a way worthy of one's powers." Some examples of the objective tests applied in deciding whether "justice" was done are whether a party had the right to counsel, whether the judge or other arbiter was prejudiced, whether there was a right to confrontation of witnesses, and whether reasonable cross-examination was permitted. There are many more. When all are met, justice has been done, virtually irrespective of the result.

In contrast to this definition of justice, "truth" is an uncertain and sometimes most illusory concept. Although truth certainly is an ultimate goal of the adversary process, it is very much a secondary one and not at all essential. One law professor critiquing my analysis here, wrote that "justice emerges from the application of the law to the truth; that is, justice is on the side of truth." Conceptually this

* However distasteful, much of the obstruction of evidence-gathering during Watergate was perfectly lawful. Another example is Missouri Senator Thomas Eagleton's brief 1974 vice-presidential candidacy, aborted when someone dug up a long-past history of shock treatments.

may be valid; practically it is of little significance, because it is a fundamental and usually even unchallengeable assumption of the adversary system that truth—whatever it may be—is best discovered by the adversary process, that is, by permitting those who have the greatest stakes in a fight to fight it out. In the final stage of their dispute, they present their evidence and arguments to an impartial arbiter, who has no independent knowledge of the merits of the controversy. So long as the contestants have had this "fair chance" at battling for the result each seeks, they will have had all that they are entitled to have. The assumption that truth has been found will not be tested except in a most extreme case—such as where a person credibly confesses to a crime for which another has been duly convicted.

"Fairness" alone—called "due process" in our federal Constitution—is the essence of our system of justice, and therefore of our adversary process. A contestant can validly and properly defeat a claim—no matter how ethically or morally "unfair" everyone knows the result to be—by showing that it is too old, and therefore invalid under a "statute of limitations"; or not in writing, and therefore barred by a "statute of frauds"; or based upon evidence which must be suppressed because obtained by an unlawful wiretap, or a search without the required judicial warrant, or improper coercion. People are willing to accept and live with such a result in an ordinary matter. They will not accept it, however, if its effect is to encourage wide consumption of a dangerous carcinogen, or to give a nuclear bomb to terrorists.

Many people appreciate that dispute resolution does not always find truth. But in the ordinary case, they are satisfied because the parties have had their "fair chance." The modern complex public interest dispute, however, does not fit into this standard adversarial mold. Here the public may well be the key contestant. Yet ordinarily it is not even represented in the struggle, and thus does not have its fair chance. Instead, the battle will usually be between a large corporation, a so-called public-interest group, and a government. (Few individuals can marshal the tremendous resources required to carry on such a fight where they are acting in their individual or personal capacities.) Each of these organizations has its own special goals and objectives. The corporation is seeking to "maximize profits." The public-interest group is committed to reflecting the views of its leadership or membership, which is inevitably a small part of the total population. The "government" represents only the particular political or social or economic policy of the officials in power. It is also peculiarly subject to influence by political constituencies, pressure groups, and the like.[6] It is only coincidental that the purpose of one of the active parties to a major socioscientific dispute will turn out to be identical to the public interest. More commonly, the true public interest lies hidden somewhere in the totality of the interests of all the direct parties, and in many other interests as well.

The public is not only unrepresented, but is also quite unable to understand the complexities of the struggle. The only thing it can observe is the "game playing" of the parties. That seems clearly designed to permit each of the participants to "win" for itself, at almost any cost—and all too frequently, at the risk of "the public be damned!" Even where the dispute has reached its final confrontation stages and judicial litigation is involved, the public recognizes the trial judge, who is the judicial arbiter, as having the same role as that of the impartial sports umpire. His assignment is to observe the battle and call shots "fair" or "foul," but to stay well out of the fray. For example, when Federal Judge John J. Sirica entered the fight during the Watergate trials in order to redirect its course along perceived public-interest lines, the public saw his legal and judicial colleagues, as well as the participants and their legal representatives, condemn him for assuming the mantle of "prosecutor." When juries are involved, trial court judges have less right to participate in the struggle between the parties. Appellate court judges are even more restricted than lower court judges in their rights to inquire independently into the facts. And facts, of course, are the critical evidence in socioscientific disputes, for they are usually the basis upon which public policy value decisions are made.

Legal tactics thus often appear to determine the outcome of socioscientific disputes, not the underlying merits of the controversy or the conduct of the parties themselves. In addition, unlike the more traditional dispute, there is no real way to test whether the outcome of the controversy is "right." The traditional dispute usually has an answer, somewhere, someplace, if only it can be found. Its final outcome (somewhat oversimplified, of course), will turn on the answers to questions such as who shot the gun, or how fast the vehicle was traveling when it entered an intersection. These can be answered. The answers may be difficult to find, especially when a large number of witnesses testify to different things. Once in a while, however, a photograph or a tape recording or some other form of positive proof will turn up. One can then decide whether the adversary system worked to find the best result. In socioscientific disputes, however, there is no such possibility of proof positive. The final decision is a quality-of-life, or value, or public-policy decision. What ultimately happens does not prove whether the decision was right or wrong. It may well have been wise to take a risk that something would not happen even if it later turns out that the undesired event occurs. If the odds against an event are 10 to 1, you may well assume the risk to gain a major benefit—but most often you will lose.

As a result of all this—lack of representation, inability to understand, fear that results are determined by "tricks" and not substance, and uncertainty that the system really works—more and more people are growing increasingly dissatisfied with the handling of socioscientific controversies. The intimate interests of the public are deeply and personally involved in these matters. People want a dispute-resolution process in which they can have confidence that the real and vital concerns involved are being properly presented and considered. These wishes are not yet being satisfied. Legislation during the last decade shows that the public is reluctantly accepting increasing layers of government regulation and control as an answer. Unfortunately, the grave danger is that too much progression along such lines will mean loss of freedom in pursuit of, and in the name of, the "public interest," but with the real public interest being even less served than before.

There is another and a better way. A large number of those who are directly involved as parties in these important socioscientific disputes genuinely believe that what they are seeking is in the public interest. When they cut out the "games," and act in responsible, straightforward fashion, they will convince the public observers of the integrity of their purposes. They will thereby maximize their chances for success.

A party's participation in a socioscientific dispute must be concerned as much with the public impact of what is said and done as it is on the other immediate parties, and on whoever the arbiter may be. This is the "rule-of-reason" approach to dispute resolution. It does not modify the useful adversary process in any paradigmatic sense.[7] It simply mandates application of that process in a way that is geared to the different kinds of problems which modern socioscientific disputes present. It maximizes the chances for success of the party employing it, and minimizes the risk of serious loss—while at the same time serving the greater public interest. It makes sense for everyone—society most of all.

The rule of reason requires that improper "tricks" and obstructive tactics must be eschewed. They may be successful in the short run, against another immediate party. Ultimately, however, the public will recognize them as game playing, even if it cannot understand the intricacies of the controversy. It will interpret them as evidencing lack of conviction in the underlying real, albeit incomprehensible, merits. It will consider them as aspects of a methodology in which the public interest is at most a remote and incidental goal. Where it can, it will penalize the party most commonly employing them.

Institutional changes in dispute resolution may well be sound. Some are considered elsewhere in this book. However, I believe that the dispute-resolution process will be much improved generally if only one of the parties to socioscientific disputes can be persuaded to adopt the rule-of-reason approach. That is because the other parties will quickly come to appreciate that the most effective way to combat the strategy is by employing precisely the same methods in opposition. When and if socioscientific battles are joined on that kind of basis—that is, with all the parties speaking for what they really believe to be the public interest, in straightforward and responsible fashion—the chances for properly finding that interest will have been enhanced. Certainly the public will have finally been satisfied that its interests are being determined in a credible manner. It may conclude that it need no longer support demands for drastic or radical changes in the system.

One "comment draft" of this manuscript came back with the notation alongside the above, "Good guys finish last!" I appreciate, of course, that the rule of reason may sound strictly utopian to some. But it is, in fact, practical and realistic. Understanding why this is so requires a more detailed analysis of the pervasiveness of the present adversary process in our society, of its consequences in major public interest matters to date, of why those charged with its administration (generally lawyers) cannot change it, and of why there is hope that one of the ever present parties to these disputes—the large corporation—may be coming to appreciate the need to introduce changes on its own. The balance of this chapter and the next two deal with these subjects.

The Adversary Process

The adversary process, emphasizing the "winning" of a "game" being played in accordance with strict Marquis of Queensberry rules, is ubiquitous in today's American culture. It characterizes the resolution of disputes not only where lawyers and courts are involved, but also disputes between individuals, organizations, governments, and—most important for the socioscientific disputes which are the subject of this book—sometimes even between scientists. Alcksandr Solzhenitsyn is right in saying that we are a "litigious society."[8]

Undoubtedly the legal profession is responsible for the careful design of the adversary process. However, its conceptual roots reach deep into the principles of individual freedom which are central to Western society. Its key assumption—that justice and truth will result if individuals are allowed to fight for their own interests—is similar to Adam Smith's underlying assumption of capitalism—that the most efficient production, distribution, and consumption of goods and services will result if individuals are permitted to pursue their own profit-maximization motivations. It is no wonder that the process has been readily welcomed in many nonlegal areas in which personal ambition and success are involved.

The process had already reached maturity in the nineteenth century. Charles Dickens' 1852 description of the never-ending case of *Jarndyce* v. *Jarndyce* in *Bleak House*, with its "trickery, evasion, procrastination, spoliation, botheration, under false pretenses of all sorts," remains the classic description of the adversary process at work.[9] Although the tale of *Bleak House* is fiction, the process described is most certainly not. Only a few years after its publication, an English commentator wrote:

> English criminal procedure does not so much seek the discovery of truth pure and simple, as the discovery of truth according to certain artificial rules. . . . The prisoner must be convicted according to the strict rules of the legal game, or not convicted at all, and that, too, however clear his guilt may be.[10]

A few years later, Roscoe Pound—one of America's greatest legal educators and philosophers—completed the description. In a brilliant and still widely circulated and quoted address delivered at the annual meeting of the American Bar Association in 1906, he said:

> The sporting theory of justice, the "instinct of giving the game fair play," . . . is so rooted in the [legal] profession in America that most of us take it for a fundamental legal tenet. . . . With us, it is not merely in full acceptance, it has been developed and its collateral possibilities have been cultivated to the furthest extent. Hence in America we take it as a matter of course that a judge should be a mere umpire, to pass upon objections and hold counsel to the rules of the game, and that the parties should fight out their own game in their own way without judicial interference. . . . The inquiry is not, What do substantive law and justice require? Instead, the inquiry is, Have the rules of the game been carried out strictly? . . .
>
> The effect of our exaggerated contentious procedure is not only to irritate parties, witnesses and jurors in particular cases, but to give to the whole community a false notion of the purpose and end of law. Hence comes, in large measure, the modern American race to beat the law. If the law is a mere game, neither the players nor the public who witness it can be expected to yield to its spirit when their interests are served by evading it. And this is doubly true in a time which requires all institutions to be economically efficient and socially useful. We need not wonder that one part of the community strain their oaths in the jury box and find verdicts against unpopular litigants in the teeth of law and evidence, while another part retain lawyers by the year to advise how to evade what to them are unintelligent and unreasonable restrictions upon necessary modes of doing business. Thus the courts, instituted to administer justice according to law, are made agents or abettors of lawlessness.[11]

We have surely learned a very great deal in the past several decades. Our "sporting" tactics are far more sophisticated than they were in 1906. But the heart of our adversary process remains the same. One recent quotation will complete the chronology begun with Charles Dickens:

> Once again this Court has been called in to arbitrate the no show and no tell discovery games engaged in by the parties to this lawsuit. I should emphasize that this is not the only game in town. The fact pattern hereinafter recited has repeatedly surfaced in other litigation during my tenure on the bench. In fact, I have often thought that if the Federal Rules of Civil Procedure were in effect in 1492 the Indians would undoubtedly have made a motion to suppress Columbus' discovery. . . .
>
> The sad part of the foregoing chronology is that the only things accomplished in this time span are the production of incomplete answers to Plaintiff's first set of interrogatories, the impregnation of my file cabinets, the generation of legal fees and the fact that I have aged a year. Or is it ten?[12]

It is important to add that the judge in the above case was not criticizing the conduct of the attorneys handling the litigation. He appreciated that the adversary system calls for counsel to act as aggressively as the client wishes and the law permits in the representation of a client. He criticized the clients only. He did not know whether the conduct was expressly mandated by the clients. However, he said that he was "mindful of the fact that often these tactics are undertaken at the behest of the party," and took "notice that many defendants instruct their attorneys to delay the litigation as much as possible, thus making the plaintiff lose money and interest in his suit." He even went out of his way to protect the indentity of the attorneys involved. When he sent his opinion out for publication, he issued two separate orders—one for plaintiff's counsel, and one for defendant's counsel—that "attorneys names not to be published at the request of Judge Robert W. Porter."

Attorneys are everywhere in modern American life—in commerce and industry as well as in government and the courts. As one popular columnist rather bitterly protested:

> The citizens of this country, we are told, support four times as many lawyers as the English, five times as many lawyers as the Germans, and ten times as many as the French.
>
> Though it seems difficult to believe, we support 20 times more lawyers than the Japanese. More lawyers live off the American public than in the rest of the world combined. We support two-thirds of all the lawyers on earth.[13]

The adversary process is similarly everywhere, even in the science which is at the heart of the socioscientific dispute.

"Adversary Science"

The adversary process is indeed so pervasive that some people appear to have forgotten that there are other ways of resolving differences. One of special relevance to socioscientific disputes is the "scientific method,"* which has been employed by science for generations. Once it was relatively pure and clean; today it is often so tainted with a desire to succeed in an objective—to "win"—that the phrase "adversary science" has begun to creep into our language.[14]

The *New Columbia Encyclopedia* describes the scientific method as follows:

> First, information, or data, is gathered by careful observation of the phenomenon being studied. On the basis of that information a preliminary generalization, or hypothesis, is formed, usually by inductive reasoning, and this in turn leads by deductive LOGIC to a number of implications that may be tested by further observations and experiments. . . . All of the activities of the scientific method are characterized by a scientific attitude, which stresses rational impartiality. MEASUREMENT plays an important role, and when possible the scientist attempts to test his theories by carefully designed and controlled experiments that will yield quantitative rather than qualitative results.[15]

There is no place in this process for personal attack. Writing in *Science*, Kenneth E. Boulding, then president of the American Association for the Advancement of Science, explained:

> The personal interest of the problem-solver, however, is not supposed to affect the solution of the problem and even though problems may involve controversy, the controversy is supposed to be settled by some kind of appeal to the facts or observations rather than to the character or interests of the disputants. Arguments ad hominem are considered very bad form in the scientific community, and there is a strong ethic of truth-telling and veracity.[16]

* The term "scientific method" is employed here in its generic sense, not as descriptive of the specific technique by which a scientific conclusion is reached. There are many "methods" in the latter sense, such as logic, observation, measurement, and experiment. They are not necessarily peculiar to science.

Scientific method requires the open dissemination of data and findings to colleagues in discussions, in scientific meetings, and in publications in scientific journals. It calls for replication by other scientists of research results, in their own laboratories, to see whether independent testing achieves the same product. It seeks "peer review," in which colleagues evaluate each other's work in other ways. Finally, there is an effort to achieve scientific "consensus." Reto Engler, also writing in *Science*, described this important latter element of "consensus," as follows:

> There is a difference between the legal and the scientific approach to a problem. The scientist can and must consider all available information; the lawyer's argument includes only those aspects beneficial to his client. At first glance this often makes the scientist appear to be indecisive and contradictious, whereas the lawyer appears uncompromisingly clear. But in order to arrive at the truth, the legal approach calls for an adversary opinion and finally for a jury or a judge to render a verdict. The scientific approach is based on argumentation among peers in order to come to a consensus opinion. A split decision is an acceptable outcome, but it does not indicate that someone is right and someone is wrong. Rather, it indicates that the available information can be interpreted in more than one way and that probably more detailed data are needed to arrive at a consensus.[17]

The differences between this "scientific method" and our "legal" or "adversarial" method are striking. Two nineteenth century English historians pointed out how the two methods are at diametrically opposite extremes of the dispute resolution spectrum in this way:

> At one of these [poles or extremes] the model is the conduct of the man of science, who is making researches and will use all appropriate methods for the solution of problems and the discovery of truth. At the other stands the umpire of our English games, who is there, not in order that he may invent tests for the powers of the two sides, but merely to see that the rules of the game are observed. It is toward the second of these ideals that our English mediæval procedure is strongly inclined. We are often reminded of the cricket match. The judges sit in court, not in order that they may discover the truth, but in order that they may answer the question, "How's that?" This passive habit seems to grow upon them as time goes on. . . . Even in a criminal cause, even when the king is prosecuting, the English judge will, if he can, play the umpire rather than the inquisitor.[18]

The following simple table will help illustrate the critical differences between scientific and legal thinking, and why the introduction of legal method into science impeaches science and scientist alike:

	Scientific	*Legal*
Method:	Nonadversarial; peaceful	Adversarial; war
Objective:	Discovery; consensus; "truth"	Win
Technique:	Analysis; testing; validation; replication; disclosure; critiquing	Persuasion; "tactic"

The two models, scientific method and legal method, are separate and distinct. Unfortunately, science does not always treat them that way. Scientists are human also. They have their own feelings and ambitions, and can be just as anxious to "win" as any lawyer or client. In addition, most are well-educated and intelligent, and are quick to learn. Some have become even more adept at employing the adversarial tools of the legal system than many attorneys. When they do so, they impair confidence in science. We then see the familiar battle of expert against expert, with the common public response often being "a plague on both your houses!"

A recent example is revealing. One of our nation's leading economists was retained to render an expert opinion in an important antitrust litigation. He had served as a member of the President's Council of Economic Advisors, and as chairperson of the economics department of one of our great universities. Clearly he had the stature and all the qualifications necessary to do the job well.

The issues in the dispute were difficult and complex, as one would expect (or he would never have been retained). He began his analysis by reviewing the relevant documents and facts, arriving at tentative and preliminary views to be tested and critiqued. After considering a number of doubts and concerns, he finally completed his work, arriving at a conclusion in support of the client's position.

Unfortunately, however, it appears that he also took it upon himself to assume the role of advocate for his client's cause and for his final conclusion as well. He participated in, and perhaps even initiated, conduct designed to obstruct inquiry by the adversary party into his preliminary doubts and concerns. He disposed of the documents he had worked with under very suspicious circumstances. When trial cross-examination came, his responses to questions concerning his documents and earlier views seemed clearly evasive. The result was to lead the trial judge to comment privately to counsel that his credibility was "destroyed on the witness stand." It is impossible to know whether the jury hearing the case understood the extremely complex issues being debated; even the trial judge may have had difficulty, for the decision was reversed by an appellate court. But the jury did appear to understand the damaging significance of the efforts to obstruct inquiry and evade the disclosure of information. Such tactics may well have contributed to a catastrophic loss for the expert's client.[19]

This "battle of experts" which attends the presentation of so many of our complex disputes is so common, and so typical, as to have become the butt of crude humor—"How much did *your* expert cost?" Two scientific experts of equal stature, reputation, and apparent competence will debate vigorously and vociferously at opposite extremes of the same issue—in the press and on television and radio, as well as in the courts and the administrative agencies. Medical experts will state certain and unqualified opposite opinions as to whether an injury was caused by an accident; toxicologists as to whether a chemical compound is a carcinogen; nuclear physicists as to whether there is radiation hazard from a nuclear plant. Each will also attack the other, not only on the data and methodology involved, but also on motivation; source of funding; civic, social, political or social view; and even sexual preference. Much of this is encouraged by the legal profession. As one article in the Harvard Law School *Record* stated:

> In a legal or regulatory dispute, the testimony of a balanced, objective scientist who scrupulously qualifies his statements to guard against overinterpretation, is likely to be almost useless to either side. Scientists with strong biases (known in advance to be most likely to suit the interests of the client) are much more likely to be sought out, if they have the requisite respectability and qualifications.[20]

This last, simple quotation explains much of what is wrong with the way in which we presently handle socioscientific disputes. The "balanced, objective scientist who scrupulously qualifies his statements to guard against overinterpretation" is precisely the kind of scientist whom science honors. The fact that he is "almost useless to either side" in a public-interest dispute simply does not make sense.

Perhaps there are times when a true battle of experts is justified. As will be discussed in chapter 6, however, the bitter differences between experts are usually regarding "value," "quality-of-life," and "public policy" matters, and not "pure" science at all. Most of the time the battles are "advocacy" struggles, because each of the experts is fighting to "win" for his cause. I have dealt for many years with scientists, friend and foe alike, both as a lawyer-advocate and as an academician. I regret to say that there are some (on my own side as well as against) who are literally red-faced in their obvious antagonism. Sometimes they violate even the common social courtesies. They may refuse to acknowledge communications and withhold important information in order to be able to surprise an adversary scientist at the embarassingly right moment. They may disclose preliminary and inadequately evaluated data to the general press, and even on radio and television, before anything has been submitted to a scientific journal for peer review. They may attack scientific colleagues in the typical *ad hominem* fashion of the trial lawyer, and engage in other adversarial extremes which negate dedication to science and the scientific method.

Intent, as earlier pointed out, is often critical in the law. It can spell the difference between prison and freedom, or between liability and innocence. Intent is ordinarily not significant in science. A laboratory experiment or an epidemiological investigation must turn out the same way irrespective of the intent of the scientist-investigator. Careful, scientific experimental controls are employed to guard against unknown bias entering the picture. Important work must be susceptible of replication by other scientists in other laboratories with different equipment. When a scientist evidences the hostile intent I have described, he negates his dedication to science. He impeaches his claim of status as objective scientist.

Scientists may not appreciate it, but this obvious hostility can be used by adversary attorneys to good advantage. The trial lawyer will write the hostile scientist-witness an admittedly self-serving letter. In the nicest possible fashion, it requests an opportunity to confer in advance of trial in order to be able to understand and properly present the facts to the tribunal "in the public interest." It goes out by certified mail, return receipt requested. Then, when it is returned, marked "signature refused," as so often happens, it is kept, still sealed. It is opened dramatically during cross-examination. It is read to the jury with compelling emphasis. The hostile scientist is hard put to explain why he or she would not even receive and read the letter—except upon the basis of the bias and hostility so vividly being demonstrated.

Scientists surely have a responsibility to society to communicate their views and conclusions to the public in the most effective, proper fashion. But if they are to preserve the integrity of their profession, and the credibility of the method by which they come to their judgments, they must do so in frank, open, and straightforward manner. When they adopt the adversarial method of the law, they impair the essence of their craft, and themselves as scientists.

For the resolution of socioscientific disputes to be achieved sensibly, we need credibility in science. Although some suggestions as to how this might be achieved are treated in chapters 6 and 7, this is essentially an assignment for science and scientists. It is one of the most important tasks they face in these final two decades of the twentieth century.

Adversary Process Risk/Benefit Analysis

Thus far in this chapter, we have seen that the fatal defect in our present method of resolving socioscientific disputes in this country results from use of the adversary process as the dispute-resolution mechanism without change. That process is everywhere, including sometimes in science. It emphasizes "justice" and "fairness"—defined in strictly procedural legal terms—over "truth" and outcome, even in matters in which the public's critical concern is result alone. What can be done about it?

The most obvious solutions are either to modify the process we use to resolve socioscientific disputes, or to modify the ways in which people and organizations employ that process. Changes in process may one day turn out to be necessary. However, the fewer structural modifications we need to introduce by fiat and mandate, the fewer uncertain and unknown risks we incur. The more we can rely on education and understanding to influence people to change their conduct in their own self-interest, the more freedom we can preserve in our society. Moreover, our present adversary process is not all bad. Its search may indeed be for "justice" rather than "truth," but the two are often enough the same to make the methodology acceptable to society in most affairs. Certainly no other process has been offered which promises better results.

Our problems are not yet so serious, or immediate, that we must introduce radical changes into our societal structure until we have at least tried a more moderate course. I believe that analysis and understanding will demonstrate that "self-serving" changes by industry in dispute-resolution tactics will take us far along the path to remedy. Many leaders of industry complain of their present poor showing in socioscientific dispute resolution. They must be made to understand that much of this is the result of the "games" their own representatives play. Prohibiting such conduct will result in decisions far more to industry's liking.

People employ the "sporting" tools of the adversary process, including its extremes, because they frequently help achieve the desired goal of "winning." Even in socioscientific disputes, delay and obstruction can prevent an undesired outcome from taking place at the wrong time. The *ad hominem* appeal to bias and prejudice can destroy an adversary. These weapons do not always backfire and blow up in the user's face. Were it otherwise, their use would have come to an end long ago.

I believe that sporting extremes are usually unsound even in ordinary cases, as a simple matter of sound dispute-resolution strategy. Even against a "strike suit," the best opposition is to fight directly on the merits of the claim. That is a strictly "judgment call," however, with which many of my trial-attorney colleagues certainly disagree. In socioscientific disputes, in contrast, I am convinced that the use of such tactics is demonstratively unwise in terms of "winning" alone—and without any need to consider what is "good" or "right." A simple risk/benefit analysis of the pluses and minuses of using sporting extremes is helpful. The plus is that sometimes these tactics help to produce the desired result—to "win." The much more powerful minus is that very often such tactics are "negative" evidence. They demonstrate lack of confidence in the merits of the case, or something to hide, or worse. On occasion, they backfire, with such disastrous consequences that the possible help such tactics might have produced seems trivial in comparison. The "benefits" of employing sporting or game extremes in socioscientific disputes are so far outweighed by the "risks," the "costs" and the other "disbenefits," as to render such tactics completely unacceptable.

The classic case demonstrating this is the General Motors–Ralph Nader controversy over the safety of the Corvair auto during the 1960s. The General Motors defense included an effort to attack Nader, by conducting surveillances, following him, and asking intimate personal questions regarding his sex life, his religious attitudes, and his drinking habits. Nothing negative was developed, but the implications of these tactics blew up in General Motors' face and damaged the company seriously. They also undoubtedly confirmed Nader as the hero and major spokesperson of the consumer movement.

> When it all came out, the consequences to General Motors were immediate and disastrous. Its corporate credibility and image, which it tried so hard to enhance, plummeted. Its Corvair, which by 1967 had by far the lowest single-car crash rate of the leading compacts (0.16 per million miles, versus 0.24 for the Valiant, 0.55 for the Falcon, and 0.56 for the Volkswagen), had so fallen in public esteem that production had to be halted in 1969. The public undoubtedly interpreted the effort to "get" something on Ralph Nader by investigating his sex life and his religious beliefs as an admission that GM knew it couldn't win on the merits of its case for Corvair safety—the only issue in the litigations insofar as the public was concerned.[21]

Firestone Tire & Rubber Co. was involved during 1978 in a widely publicized dispute regarding the safety of its "500" brand, passenger, steel-belted radial tires. It finally settled with the National Highway Traffic Safety Administration (NHTSA), by agreeing to a tire recall which *Business Week* magazine termed "so large it almost defies perspective." About 13.5 million tires were involved, at a cost estimated at more than $200 million.[22] The *Legal Times of Washington*, a weekly lawyers' newspaper, discussed the case in a lengthy article entitled, "Firestone Case: A Lesson for Lawyers." It reported:

> But students of the law can also benefit from Firestone's troubles. The company elected to treat the problem strictly as a legal headache; its early tactics involved litigation, delay, contentious foot-dragging and appeal. . . . What was Firestone trying to conceal? Why was [counsel] so concerned? . . . In a lengthy, contentious answer drafted by [counsel], Firestone objected on grounds that the order was unreasonably burdensome, not sufficiently specific and exceeded NHTSA's authority. . . . U.S. District Judge Thomas A. Flannery rejected each of Firestone's contentions in an Aug. 15 ruling, criticizing [counsel] for tardiness and supporting NHTSA's authority. He said Firestone would not be unduly burdened by producing information "which it has known for six months was being sought."[23]

Business Week commented that the damage done by the adverse publicity was such that "Restoration of Firestone's reputation in the retail market is likely to cost more than the recalls," and concluded:

> Firestone Chairman Richard A. Riley still maintains that the tires are safe, and he says Firestone has been forced by adverse publicity to capitulate. That position may serve Firestone as it defends itself against about 150 lawsuits now pending over the 500. But when asked at a press conference if he was satisfied with the terms of the recall, Riley replied, "I'm not satisfied with losing anything, any time." That attitude would appear to be the same attitude with which Firestone entered the dispute.[24]

During that same year, 1978, Ford lost a number of multimillion-dollar, automotive safety cases, including one $4 million and one *$125 million* "punitive-damage" award ("punitive damages" can be equated to findings of willfull misconduct), and was criminally indicted to boot. Not all were sustained—the $125 million punitive award was reduced promptly by a judge to $3.5 million, and a jury found Ford not guilty of the criminal charges—but the losses clearly reflected judge and jury dissatisfaction with Ford conduct. In an article entitled "How Ford Stalled the Pinto Litigation," discussing Ford dispute-resolution tactics in some of these cases, the *American Lawyer*, another weekly lawyers' magazine, wrote:

> According to plaintiffs' counsel around the country, Ford's discovery behavior . . . was emblematic, setting a pattern of delay and nondisclosure that the company has continued through all its fuel tank litiga-

tion. "The way they cold-bloodedly say the documents aren't there—I have never seen any defendant behave like this before," says Hare, who has had 20 years' experience in products liability. "Ford is absolutely unique."

Like Hare, other plaintiffs' lawyers have been forced to go to trial ill-equipped, or have been repeatedly hamstrung through years of pre-trial maneuvering. Some crucial documents haven't turned up until after a trial, often after a plaintiff's case has been lost. Ford has been cited in at least five appellate court decisions for having obstructed discovery by giving false answers to interrogatories, and by hiding damaging documents.[25]

When IBM settled the antitrust suit which its competitor, Control Data Corporation, had brought against it, one of the conditions of settlement was that Control Data destroy a computerized data base it had developed. This had made possible rapid access to a very large number of documents to be used at trial. Both the Justice Department and another IBM competitor, Telex Corporation, had been relying on continued access to Control Data assistance for the preparation of their own cases against IBM. They protested bitterly when they finally learned about the destruction. But it was too late—the process had already been carried out, on a rushed and clandestine basis, before anyone, including the courts in which the cases were pending, could find out about it.

Although everyone fudged why destruction was so important, the clear and obvious exclusive purpose was to prevent others from being able to use the data base.

But despite its apparent ethical propriety, the public understandably interprets such conduct as evidencing an intent to hide something damaging, and to obstruct the efforts of third parties to assert their claims. It views such actions as totally inconsistent with a corporate claim that its conduct is proper, that it has nothing to fear from the facts, that it wants its day in court, and that when all the facts are known its actions will be found to have been proper and in the public interest. . . .

The destruction quickly became major news in the general as well as the commercial and trade press. The damage to IBM's reputation and credibility, which it also works so hard to develop, was tremendous and apparent. . . .[26]

A jury in an insurance case, which involved only $45,600 in compensation, awarded a plaintiff $5 million in punitive damages against

Mutual of Omaha Insurance Co. Although the trial judge promptly cut this award in half as clearly excessive, he said that the conduct of Mutual of Omaha witnesses "could well have given the jury the indication of a deliberate cover-up."[27] The *New York Times*, in an editorial commenting on the bitter and lengthy Reserve Mining Lake Superior asbestos-like fiber pollution case, had this to say in defense of the clearly improper conduct of the trial judge:

> But there are moments in the life of the law when exasperation at pettifogging and delay is an understandable response. Judge Lord's indignation at Reserve's outrageous tactics has served at least to pierce the crust of public apathy and to make the public health menace real and understandable.[28]

In fairness, it must be pointed out that no independent investigations have been conducted regarding the above cases. The socioscientific dispute is still a relatively new one in our society, and no adequate empirical analyses of the use of adversarial extremes in such disputes have yet been published. Information of this kind is also extremely difficult to come by. Parties employing these tactics are understandably reluctant to disclose the intimacies of their conduct, especially while the matters involved, or related matters, are still pending. Firestone and Ford, and their counsel, have on occasion protested the accuracy of some of the published reports about these cases. Although I know of no inaccuracies in the statements quoted above, it may be that they will take exception to something repeated here as well.

For present purposes, however, the important facts are those with regard to the public perception of what happened in these cases, perhaps almost irrespective of the actual events themselves. These cases are valid evidence of the public response to the use of perceived adversarial extremes. They show that serious financial losses and even criminal and quasi-criminal penalties may result from adverse inferences drawn from dispute-resolution tactics. The question of whether a tire or an automobile is societally unacceptable becomes secondary or lost. Whatever the underlying truth, these reports support the conclusion that the use of adversarial extremes in socioscientific disputes is unwise and even dangerous.

Throughout this book, I will use a number of different terms to characterize certain kinds of dispute-resolution tactics. These include "game," "sporting," "obstructive," "trick," "cute," and "adversarial extreme." Thus far, I have referred only to the definitions of Roscoe Pound and the earlier English authorities quoted above. As one law professor put it, "Gamesmanship is only on the side of victory." I am an advocate of the adversary process. I believe that lawyers, and parties, should be permitted to engage in conduct designed to achieve their objectives, within limits to be discussed in chapters 3 and 4. My definition of these terms is therefore more limited.* I use them to identify lawful conduct which, if fully disclosed in all its intimacy, would be interpreted by a disinterested onlooker as somehow contrary to asserted, or implied, purpose or position. It would be embarassing.

This definition is keyed to the critical problem posed by the employment of adversarial extremes in socioscientific disputes. That is that the public may perceive them as inconsistent with a stated position. The public does not understand the scientific merits of the controversy, and therefore decides who is right and who is wrong on the basis of this collateral evidence of tactics. It thinks it can understand such tactics as evidencing consciousness of culpability. It is much the same as evidence of a defendant fleeing the scene of a crime.

Delay, as such, is by no means necessarily antisocial, or contrary to the public interest, or "sporting." There is nothing even remotely suspect about a defendant rising in court in a criminal case, and openly requesting delay until the passion of the moment has abated and an impartial jury can be selected. A corporation can do exactly the same thing without embarassment if, for example, time is needed to complete laboratory research to determine whether a questioned chemical is a carcinogen. Delay becomes an adversarial extreme, however, when the corporation asserts its innocence, demands immediate adjudication, and then initiates "exhaustive discovery" solely in order to bleed the other side, and to force opposition counsel to "turn over a couple of times"—common tactics openly recommended by one prominent trial lawyer, whose remarks are discussed in the next chapter.

* One law school colleague has suggested that it is improper to cite Pound's arguments against sporting tactics, because my definition is different. But Pound's definition certainly includes all that I term "sporting," and much more. His arguments thus apply *a fortiori*.

It warrants emphasis that *both* definitions apply only to conduct which is legal and technically ethical and proper—however unsound or unwise it may be socially or otherwise.

Why are sporting tactics as I have defined them, so much more dangerous in socioscientific than in other disputes? First, because the public is so much more intimately concerned with result rather than process. It is unhappy with procedures which seem designed to avoid an outcome in the true public interest. Second, because the public is unable to understand the complexities of the science involved. It has no adequate alternative except to make judgments on the basis of collateral evidence—delay, obstruction, and the like—which it *can* understand. Third, because socioscientific disputes will not go away. They have a way of going on and on and on, for years, as new problems emerge, as new evidence and information develop, and as conditions change. One may be able to "get away" with a cute "trick" in a relatively brief negligence case, or criminal prosecution, or marital dispute, and then close the books once and for all. In a socioscientific dispute, however, that little tactic can show up years hence, in some other context or connection, and after one's adversary has had a good chance to decide what to do. As Macauley Whiting, chairman of the industry caucus of the National Coal Policy Project (NCPP), (discussed in chapter 7), has observed, "Winning the battle in litigation gets a trade-off of losing the war of credibility."

However one regards sporting extremes generally, they can be self-defeating when employed by those carrying the societal burden of persuasion in socioscientific disputes. They create risks of harm far greater than any benefits they may produce. Without regard to their potential for bringing on much greater government regulation and control than anything we have yet seen (to be discussed in chapter 5), in my opinion they have already seriously damaged corporate credibility. Surveys show that 80 percent of the public believe that a few large companies hold too much power; 60 percent believe that large companies ought to be broken up; 71 percent think businessmen act mainly out of self-interest; only 23 percent have "a great deal of confidence" in the people running major companies; and 87 percent of students—those who one day will lead our country—believe that people in business are concerned too much with making a profit and not enough with public responsibility.[29]

The view of the legal profession is almost as bad. It is reflected in the analyses and conclusions of the recent National Commission for the Review of Antitrust Laws and Procedures (NCRALP); studies by the American Bar Foundation, the research arm of the American Bar Association; in an increasing number of comments by popular columnists and national magazines; and even in attacks by the President of the United States and other leading public officials.[30] I am convinced that their own adversarial extremes have been important contributors to these low esteems in which both corporate and legal societies are held.

Perhaps the most objective evidence of public disapproval and disenchantment is the poor showing of industry in the resolution of socioscientific disputes in the courts and in the administrative agencies. Corporate parties and their attorneys are proud of the tremendous scientific and other resources they marshal in support of their contentions. Often they face poorly funded and inadequately staffed environmental or consumer organizations as key opponents. Yet industry leaders correctly complain about their dismal batting record. The Environmental Defense Fund (EDF), the National Resources Defense Counsel (NRDC), and similar organizations, have virtually brought new nuclear-plant licensing to a standstill, and have an astounding list of court and agency "wins" in the lawbooks.[31] Unfortunately, far too many of these decisions have been based on tactics, and not on the merits of the case. The public deserves better.

Chapters 1 and 2 have sought to define the modern socioscientific dispute, and to explain what is wrong with present dispute-resolution methodology and why. We turn now to a discussion of the roles of the key participants in the process—lawyers, corporations, and scientists. This, in turn, will suggest how the system can be improved, and by whom.

LEGAL RESPONSIBILITY

Attorneys are commonly involved in the resolution of socioscientific disputes. When litigation is reached, they are *always* participants. The law charges them with obligations as "officers of the court." If the use of sporting extremes results in decisions which are contrary to the public interest, why should not they be required to institute the needed changes? Amendments to professional codes of ethics and disciplinary rules might render such conduct unprofessional, and thereby eliminate at least the most serious aspects of the problem.

This is an effective approach, where the public interest is clear. Indeed, certain adversarial extremes have long been prohibited to attorneys. These include conduct which is in violation of general law, such as perjury, subornation, and obstruction of justice. They also include other conduct which is inconsistent with responsible decision-making, such as delay tactics undertaken exclusively for the purposes of delay, or character assassination for the sole purpose of harassment. It may be difficult to prove such delay or character assassination but they are unquestionably unethical.

In most cases, however, the public interest is not at all clear. It may be the subject of very considerable difference of opinion. Adversarial tactics, including extremes, almost always serve some proper purpose. Extensive discovery demands do develop information, even if they also cause great delays. Cross-examination as to motivation does tend to impeach a witness, even if it also leads to substantial embarassment. To compel an attorney to choose between these tactics and some vague or ill-defined "public interest," would cast him in the role of judge or keeper of the public interest. To permit inquiry into the mental process of the attorney when he employs such tactics would impair, and perhaps even destroy, his ability to act as effective advocate for his client. "Intent" is an uncertain enough matter in normal affairs. The intent of an adversarial advocate is necessarily far more suspect. It is not a sound concept upon which to base an important societal process.

There are many things lawyers and others can do to improve socio-scientific dispute resolution. But if the adversarial paradigm is to work properly, the attorney's primary role as advocate must be preserved. An understanding of that role is an essential first step in the improvement process.

The Lawyer's Advocacy Role

American society places an extremely high value on the lawyer's role as representative of his client, and as advocate of the client's cause. It imposes a specific duty on the attorney to serve the client with utmost devotion and fealty. It has always been most reluctant to place unnecessary impediments in the path of the attorney's use of weapons on the client's behalf. The fear is that the ability to represent in effective fashion might be somehow impaired.

This societal reluctance to interfere is especially strong where a proposed impediment is also a subject of the dispute itself. That would be the case were we to require that some inadequately defined "public interest"—about which reasonable people might differ—limit the adversarial tactics to be employed in a dispute seeking to ascertain what that very same public interest may be. For this reason, society has not even laid down precise rules for how to decide what adversarial tactics are to be employed, except in the clear cases of the unlawful or unethical conduct referred to above. The lawyer's sole legal obligation is to "win" for his client, employing tactics within the rules. There is no well-defined or precise standard for which tactics he should pursue, and which he should reject.

Assume a nuclear plant licensing proceeding. A formal request has been made for the production of certain documents. The lawyer representing the corporation applying for the license has in his possession one document clearly coming within the request, but also clearly exempt from production by virtue of attorney-client privilege. Should he waive exemption, and produce the document? Should his decision turn on whether the document is favorable or damaging to his client's case?

Certainly the usual client, including even the large corporation, wants an attorney who is a tough infighter, and who knows all the tricks of the trade. He is not seeking an advocate who will preach morality or ethics, or insist on waiving technical advantage because it is not "nice." Often the highest praise a client can pay his lawyer is that "he's a real, no good, S.O.B!"[1] Of course, attorneys know this and want to satisfy their clients. Attorneys are just as human as anyone else. They need food and shelter, have families to support, jockey with partners in their firms and colleagues in their corporations for position, are ambitious, and want not only to survive but to succeed. There are some important adversarial relations between attorney and client (such as fee setting or charging, and case settlements), but in general most of the attorney's inclinations and instincts are served by serving the client. To ask lawyers also to serve a "public interest" which is uncertain and undefined, and about which different people may differ, is far more than can be expected.

Moreover, in the ordinary case, it is probable that the client will know the "public interest" better than the lawyer. The attorney is commonly retained to handle a specific matter or matters, not the whole of the client's business. Even the corporate general counsel who is employed full time by his corporate client often has a quite limited portfolio. He is not involved in the overall public-interest management planning and decision-making of the organization. The attorney may honestly, and even reasonably, conclude that the tactics in which he proposes to engage in some specific matter are well justified. He may simply not appreciate the real possibility that they can come home to roost at some remote time and place, in a very different matter which he is not handling, or about which he knows nothing at all.

Even though it is nowhere specifically spelled out, in the context of the traditional attorney-client relationship the lawyer must make his decision about tactics on the basis of his client's set of "public-interest" values, not his own. In the hypothetical nuclear licensing case above, he must let his client decide whether or not to waive attorney-client privilege. This is because the attorney is a representative, not a principal, with no legal right to an independent "public-interest" ethic of his own. He can no more give away protected client information because he thinks society would be better off with disclosure, than he can refuse to defend a court-assigned client charged with crime because he thinks society would be better off with his client in prison.

Sometimes, especially in a criminal prosecution, a client is uneducated, or less than fully competent. Then the attorney may have to decide on the basis of his assumption of what the client would or should want if able to understand or decide.[2] In socioscientific disputes, however, the client is almost always a sophisticated, or knowledgeable, organization or business person. There, the client alone should make the choice between possible alternative courses of action.[3]

It is surprising how few people understand this limitation on the right of the attorney in our system. In mid-1977, I addressed a conference on "Ethical Issues in Occupational Medicine," cosponsored by the New York Academy of Medicine and the National Institute for Occupational Safety and Health (NIOSH).[4] I tried to explain the lawyer's dilemma in public-interest cases, where he was caught between an obligation to a client and his own perceived obligation to society. I argued that the client, and the scientist-witness, had obligations to define the public interest for the attorney, as discussed in chapters 4 and 6 of this book. In the sometimes heated discussion period which followed my address, one of the scientists present repeatedly expressed the view of many of his colleagues. He rejected the adversarial model, and said the lawyer should be "nonadversary." Over and over, in response to my comments, he said that the lawyer's obligation should be to "seek the truth." He would not accept my response that this is not the lawyer's job in our society.

U.S. Supreme Court Justice Potter Stewart has endorsed this allocation of public-interest responsibility between attorney and client in commercial practice. In a 1975 address to the American Bar Association's (ABA) Section of Corporation, Banking and Business Law, he stated:

> But beyond these and a few other self-evident precepts of decency and common sense, a good case can be made, I think, for the proposition that the ethics of the business lawyer are indeed, and perhaps should be, no more than the morals of the market place. The first rule of the business lawyer is to provide his total ability and effort to his client. . . .
>
> But aside from the inescapable responsibility that his profession places upon every lawyer to act as a wholly honorable and trustworthy person and a good and law abiding citizen, is there any way in which a business lawyer can better serve the public interest than by giving the best possible legal advice to his clients? Is it the duty of a lawyer, by contrast, to try to impose upon his clients his own notions of social or political or economic morality? Is it indeed even "ethical" for him to try to impose his own system of moral priorities and social values on his clients' business decisions in the guise of neutral legal advice?[5]

The practice of the American bar is clearly in accordance with Mr. Justice Stewart's analysis. Attorneys routinely and regularly change their positions to accord with what they are advised are, or what they perceive to be, the "public-interest" positions of different clients. The public sometimes treats this as reflecting lawyer disingenuity or worse, but that is not what it is at all. It is simply client representation. Thus, when representing the United States, one of the Justice Department's chief antitrust enforcement attorneys condemned the dilatory tactics of so many of his adversarial defense-counsel opponents in the following strong language:

> While not intending to exhaust the means available for intentional delay of investigations, I would point to the following types of conduct which may be evidence of bad faith on the part of counsel: Promises voluntarily to comply with document demands, in lieu of compliance with subpoenas, which are not promptly kept and were not intended to be kept; motions to quash subpoenas, the bases of which are boilerplate objections which have already been repeatedly rejected by the Commission; request after request to extend the deadline date for compliance with the subpoena, while no reasonable effort has been made in the document search. Litigation is also fraught with opportunities for inexcusable delay, e.g., making unnecessary motions to extend dead-

> lines, or to change pretrial conference dates; and asking for subpoenas to be issued to third parties, knowing full well that enforcement of them will probably be tied up in federal court, while little valuable evidence is expected to be forthcoming.[6]

Three years later, when this same attorney had left government service and had commenced to engage in private practice with one of the major corporate-law firms, he outlined his recommendations for the defense of such antitrust suits before over two hundred participants at the Tenth Annual New England Antitrust Conference. *Antitrust and Trade Regulation Reports* (*ATRR*), a widely circulated lawyers' reporting service, summarized his comments in what it described as a "frank discussion of tactical considerations for attorneys representing corporate clients in the prosecution of government and private antitrust law suits," as follows:

> In monopolization suits, [he] recommended that a defendant take exhaustive discovery, particularly if it has an advantage over the plaintiff in terms of resources. . . . [He] also suggested that any defendant show the plaintiff that it is not costless to sue. Thus, a defendant should counterclaim. . . . [He] bluntly suggested that private plaintiffs look at their pocketbooks rather than the so-called "public interest," so defendants should make plaintiffs worry about their pocketbooks. . . . [With regard to a government suit, he] suggested that defendants take at least enough time for the government staff to "turn over a couple of times." . . . In sum, he stated, settle strong cases and try the weak cases, always while delaying the Government.[7]

Another former senior government antitrust enforcement official, also now in private practice with a major law firm generally representing corporate parties, outlined the alternatives for consideration when a "smoking gun" document is found. Speaking before the American Bar Association's National Institute on "Preventive Antitrust," the same legal service reports that he included among the options:

> The corporation . . . can choose to remove the "smoking gun" document to a safer place with the hope that the document will be destroyed under the company's control program at the proper time. Other options listed . . . included *deep-sixing the document*. . . .(italics added)[8]

A government criminal prosecutor during the early 1960s had been charged by his adversary defense counsel with engaging in a number of adversarial extremes, to little avail. Over a decade later, when this prosecutor had become a distinguished law professor and could look back on his earlier conduct in the light of the academician rather than the trial attorney, he described what he had done, and concluded as follows:

> Having spoken of the Department [of Justice], of [United States Attorney] Morgenthau, and of [U.S. Attorney General] Kennedy, let me now speak of myself. I read over this narrative and I am not proud. When Morgenthau said, "The Department wants Cohn," I replied, with enthusiasm, "I'll get him." When Morgenthau showed me Cohn's income-tax return, I did not ask him whether it was proper that he have it. When I thought it necessary to obtain the Brandel records, I made the promise that would get me the records, careless of the legality of my conduct in Switzerland, where I then was; and later, when it suited my purpose, I broke the promise. When I wanted to know what Foley would say to Hagenbach, I bugged Hagenbach's room, securing a consent from Hagenbach that can hardly be regarded as freely given, and closing my mind to the affront to Foley's privacy. It was the power of power. If I possibly could, I was going to be the one to do the job the Department wanted done. Not once did I stop to think what it was a Department of. (footnote omitted)[9]

The principle that "the end does not justify the means" is in many respects fundamental to our American legal structure. Indeed, the adversary process emphasizes "means," or method, sometimes almost to the apparent exclusion of results. In other respects, however, the client-representational model we have been discussing seems to look the other way. It encourages going as far to the border of legality as possible in the employment of "means" so as to achieve the result desired by the client. Outspoken criminal defense counsel William M. Kuntsler described his advocacy philosophy in less circumspect, and therefore far more revealing language, as follows:

> Take the difference between myself and a Nixon or a Mitchell. They argued that anything was proper so long as it ensured the re-election of the President, which they saw as being in the best interests of the country. And I would argue the same—namely, that the end justifies the means. Only my *goal* is different. Daniel Berrigan and company, whom I defended in the Catonsville Nine trial, were charged with burning Government records; at the same time, though we didn't know it, John Dean and L. Patrick Gray were also destroying records. I see Dean's and Gray's work as bad because the goal was bad, while to me Berrigan's act was OK. It's just a question of goals.[10]

By now, the lay person who is not privy to legal advocacy in this country may be shocked at what I have reported. Let me make clear that with the possible exception of a technical violation of Swiss secrecy law—and assuming, of course, that the speakers were discussing advocacy and not outright obstruction of justice—I know of no one who has suggested that any of this conduct is legally questionable, much less unprofessional or unlawful. The attorneys involved spoke voluntarily and without compulsion and, with the exception of the former prosecutor, with no concern that their colleagues might think the less of them. Indeed, in three of the instances, professional legal colleagues were being advised as to how they should conduct themselves. The tactics described may well be unwise and unsound, and likely in the long run to hurt their clients' cases. But they are *not* improper.

A change in position does not reflect on an attorney. This is because it is the client's position which is being advanced. If the clients are not the same, there is no inconsistency. The issue is very different, of course, if inconsistent public-interest positions are advanced on behalf of the same client, even by different attorneys. Then the client's credibility and integrity become suspect. This is a serious concern in public-interest matters. The problems it presents, and how to control them, are underlying issues in the next two chapters.

I do not mean to suggest that attorneys have no obligation to consider the implications of adversarial extremes in socioscientific disputes. Such disputes are new and different. It is just simple, good, sound legal practice for an attorney handling one to be alert to the possibility of adversarial tactics going awry in unanticipated ways. In ordinary matters, frequently dispute-resolution tactics are not discussed with clients.[11] In socioscientific matters, however, caution suggests that doubtful or possibly questionable tactics should be discussed with, and evaluated by, the client. Certainly, where an attorney does recognize the potential negative consequences of employing certain adversarial extremes, he has a responsibility not only to advise the client, but to be sure that the client understands the lawyer's concerns and specifically authorizes what is to be done.

If he chooses, an attorney may, in addition, explain to the client what he regards to be the "public interest," which is to be considered as contrary to the employment of the questioned tactics. However, he has no right to suggest or imply that this opinion involves expert legal advice, concerning which he has special qualifications superior to those of the client. He may even withdraw as counsel if he feels strongly enough about the public interest involved, provided that to do so does not somehow result in unreasonable prejudice to the client's interests (such as by leaving the client unrepresented on the eve of trial). Further he may not go. He may not assert his own perception of the public interest as superior to that of the client and proceed to act on that basis, contrary to the client's instructions, or contrary to what he knows the client's instructions would have been had the client been asked.

This lack of a legal "public-interest" ethic has not been of public concern in ordinary matters. There are notorious exceptions of course, but most thinking people appreciate that "guilt" in a criminal case is a legal conclusion, which only a judge and jury can make after all the rules have been properly followed. A defendant is presumed innocent until proved guilty in accordance with those rules. The attorney who refuses to assert some "technical" or unsavory defense on a client's behalf, such as that a confession was unlawfully coerced or evidence unlawfully seized, is violating his client's constitutional right to counsel. Nor is the lawyer blamed in the civil case, even when the "just" claim of a poor, widowed, "little old lady" is defeated by the technical bar of a statute of limitations or of frauds. Once again, to deny such a defense to the other party is to deny that party the right of representation by counsel.

The public concern is different in socioscientific disputes, however, There, nonlawyers—including even arbiters who are scientists and who are not privy to legal ways—often express irritation and anger at attorney tactics which seem designed to "win" rather than to find the public interest. The dissatisfaction with the process is justified, but the "fault"—if it be that—is not that of the lawyer. It is that we have tried to fit a controversy which is new and different into the framework of traditional dispute resolution, without change. The emphasis, therefore, continues to be upon "winning," rather than upon the public interest. Like it or not, the lawyer alone cannot change this. To do so would not only be contrary to all of his tradition, but would prevent him from performing the other vital functions which society demands of him. He can be helpful, but we must look to his client and elsewhere for any complete answer.

Legal Ethics

Lawyers are representatives, not principals. They may not employ their own legal "public-interest" ethic to determine the manner in which the socioscientific disputes of their clients should be resolved. They do have a responsibility, however, to do what they can to improve the present sorry state of such resolution. This section discusses that "legal responsibility." Coupled with "client (corporate) responsibility" and "witness (science and scientist) responsibility," an overall approach may be developed which will serve the public interest in optimum fashion.

Lawyers' codes of professional responsibility and judicial conduct—with their canons of ethics, ethical considerations, and disciplinary rules—are lengthy documents. With the accompanying large body of case and opinion law, they constitute as comprehensive a set of ethical guidelines as those of any profession. There are elaborate grievance and enforcement procedures to match.

Unfortunately, most of these standards and procedures are grounded in the highly personalized individual or small law firm practice of a much earlier day. My father was a "single practitioner" in the law. He dealt with his clients on a face-to-face basis and knew most of them intimately. He knew how they felt, what they wanted, and what was "right" or "good" for them. He could find most of the answers to questions about "ethics" or "professionalism" by quick glances at the canons of ethics. He rarely had to, however. The answers were usually clear enough.

Times have changed. World War II sparked enormous shifts in the impact of science on society, and on the role of the corporation in society. It also dramatically changed the way in which we practice law. Even in 1948, right after the war, my father could not understand how I could begin my legal career with one of the new "giant" law firms. Yet I was only the forty-first lawyer on its staff. Today there are law firms with offices all over the country and throughout the world, many "paralegals," and huge nonlegal "support" staffs. One has over five hundred attorneys. There is no apparent limit in sight.[12]

Inevitably, some of the attorneys in these large firms will differ with their colleagues regarding ethical issues. It is almost an equal certainty that others, attorneys and nonattorneys alike, will violate client instructions, ignore ethical mandates, and even commit crimes. Internal supervision and "policing" are needed to insure compliance. Most firms do have systems for discovering potential client "conflicts" before new matters are accepted. Apart from this, however, there is very little enforcement activity. It is sorely needed.

Most clients of large law firms are large organizations, with financially important commercial problems. That kind of practice is highly "depersonalized." The associated ethical and professional problems are accordingly mainly "institutional" and "systems" problems, not individual and human ones. Some of these new problems will be considered later in this chapter and the next. They include questions such as: "Just who is the client when a lawyer is dealing with a large organization or a government?" "What is the obligation of the lawyer to the several 'publics' which are involved in commercial issues, such as the 'potential stockholder' public, or the 'consumer' public?" "Is it possible to define more precisely the murky line dividing propriety from impropriety in major commercial matters?" "Can the lawyer's responsibility of 'competence' be better defined, so as to reduce problems resulting from lack of knowledge or training?" [13]

The bar is, of course, familiar with the need to modernize legal ethics in light of these developments in recent years. Much work has already been done and more is under way.[14] The process is moving much too slowly, however. It must be speeded up. Common-law lawyers are trained in the doctrine of *stare decisis*—historical precedent is often controlling. They are inherently conservative and reluctant to change as a result. Difficult as it may be, we must overcome this slow pace. Society is moving too fast. It may leave us behind.

One of my law school colleagues fears that concentration on improving the ethics of the corporate attorney can have its own dangerous side effects. Special "white shoe" ethics for business lawyers can widen the already too-great separation of the "corporate" bar from the "individual" bar, which continues to engage in the personal practice of the past. The result can be greater disharmony and conflict, and alienation of those relegated to what some lawyers may consider to be an "inferior" class.

There is some justification for this concern, of course. We are all human. But the warning should serve only as caution and instruction in revision, not as sanction for less than optimum ethics and professionalism in any area of law practice. "White collar" or "corporate" practice is no better—and no worse—than "criminal," "negligence," "marital," "landlord-tenant," or any other kind of individual practice. It is just different. There is sound precedent for applying different ethical rules in different kinds of law practice.[15] Changing ethics in one area need not imply that the new will be "better ethics" than in other areas—they will simply be different ones. Moreover, if ethics are improved in one area, the benefits should spill over to improve all others.

We must not ignore one kind of law practice for fear of widening the disparity. We must work for improvement throughout, so as to narrow it.

Improved Ethical Enforcement

It is important that ethical standards be modernized, so as to reflect changes in the practice of law. It is also important that existing ethical standards be enforced properly. There is substantial evidence that present enforcement procedures are inadequate. Insofar as the practice of "business law" is concerned, the persons charged with enforcement responsibility are often the same ones who engage in such misconduct and their colleagues. The trial lawyer who headed the Roscoe Pound American Trial Lawyers Foundation year-long study of legal ethics codes, put it graphically and accurately, when he said: "Putting lawyers in charge of their own ethics is like putting Dracula in charge of the blood bank." [16]

The ABA's Antitrust Section conducted a study of ethical conduct in antitrust cases. It concluded that "sixty-nine percent of the respondents had encountered unethical practices while engaged in complex civil antitrust litigation. The most common practices were destruction of evidence and 'tampering' with witnesses' responses." [17] On the basis of this kind of information, the NCRALP Special Advisory Panel on Ethical Issues in Complex Antitrust Litigation concluded:

> The Panel believes that there exists a fairly limited but important category of actions that occur from time to time in antitrust litigation which are clearly unethical. Deliberate destruction of evidence, deliberate concealment of evidence, and deliberate "tampering" with wit-

> nesses in order to try to prevent their testimony from being truthful, all come within this category. The Panel believes that such practices are clearly subject to severe judicial sanction, and, in some instances, are violations of criminal law. The Panel also believes, however, that such practices are unethical under existing Codes of Professional Responsibility, and that in addition to, or in lieu of, judicial sanctions, grievance proceedings should be initiated against attorneys who are believed to have engaged in such practices. . . . [A] general problem of attitude exists among most lawyers engaged in litigation, whereby the emphasis on winning at all costs can tend to make questions of ethics appear irrelevant in a litigation context and whereby conduct on behalf of a client can result in serious adverse consequences to the efficient and effective administration of justice and thereby constitute an abuse of the adversary system.[18]

Little of this is news to lawyers. Indeed, the conduct is widespread and almost notorious. One attorney, seeking clemency for another who had pleaded guilty to a criminal charge of lying in a business case, expressed the feelings of all—including, he presumptively hoped, the judge— with the words, "There, but for the Grace of God, go I."[19] What is being done to correct these problems?

Former Supreme Court Justice Tom C. Clark chaired a special ABA panel, which conducted a three-year nationwide study of lawyer discipline. The committee report, published in 1970, found "the existence of a scandalous situation. With few exceptions, the prevailing attitude of lawyers toward disciplinary enforcement ranges from apathy to outright hostility."[20] Three years later, little had changed. *Business Week*, in an unsigned comment entitled "A Crisis of Double Standards at the Bar," stated with regard to the role of lawyers in Watergate:

> Thanks in no small part to the sorry failure of the bar to put teeth into its own enforcement machinery, its "ethics" have permitted this nation to come to the brink of constitutional disaster with the unedifying spectacle of the President, a lawyer, almost refusing to obey the lawful mandate of a federal court. . . .
>
> If the bar is to preserve its tradition of self-regulation, it is essential that the American Bar Assn. and the state bars declare war on the double standard that has permitted some lawyers to bring the nation to this crisis.[21]

New York City probably still has the largest commercial law practice in the country. The experience there is revealing. Its leading bar association is the Association of the Bar of the City of New York. That group has often been attacked as "dominated by Manhattan Brahmins."[22] In 1976, a special committee of that association, dubbed the "Silverman committee," conducted an inquiry into its grievance procedures. Part of the committee's conclusion was:

> The limited reach of the [disciplinary] system feeds public distrust of self-discipline and the skepticism of those lawyers who doubt that big-firm, "big-business" practitioners, the colleagues of those who play important roles within the bar-operated grievance systems, are likely to be disciplined.[23]

The Silverman committee recommended sweeping changes in the system. Once again, nothing happened. Two years later in a follow-up, Tom Goldstein, then legal affairs reporter for the *New York Times*, wrote: "Since that report, in spite of ample opportunities, no lawyer from a prominent Manhattan firm has been publicly disciplined, except after a criminal conviction."[24]

The "double standard" by which some lawyers are prosecuted, and some not, is illustrated in the handling of one complaint to the city bar's Grievance Committee. The complaint related to alleged attorney misconduct during a long-running socioscientific litigation. It involved prominent bar leaders, who were also important partners in a major business-practice law firm. Their conduct had been ruled to be improper by one federal judge and questioned by another.

The Grievance Committee responded to the initial complaint, saying that it would not act until after the litigation had been concluded. Thereafter, the complainant continued to register protests, at least yearly. Sometimes the committee simply ignored the renewed complaints. On other occasions, it either repeated its initial response that action would not be taken until the litigation was finally over, or said that it was not allowed by law to discuss the matter at all.

In the meantime, William Kuntsler—neither a "Manhattan Brahmin" nor a city bar leader—was convicted of contempt. The same Grievance Committee before which the business lawyer complaint continued pending, decided to proceed against him, despite the pendency of his appeal. It explained publicly, contrary to its sometimes expressed policy of not speaking at all for legal reasons, that "it was decided against waiting until Mr. Kuntsler finished appealing his contempt conviction because 'the case has been dragging on for four years and may go on another five or six.'"[25]

Reading this, the patient socioscientific complainant expressed delight to the Grievance Committee that the policy had been changed. Once again, the request for an investigation was renewed. Once again, however, the committee refused to act. Instead, the association decided not to proceed against Kuntsler either![26]

The problem of enforcing legal ethics is not a simple one, and there are few easy answers. A major accelerated effort clearly is required. If bar associations, courts, legislatures, lawyers, and the public will all recognize that fact, real improvements need not be very far away.

New Approaches

Most people solve most problems without lawyers or courts. Socioscientific disputes at present are the most difficult of all, so that perhaps more legal involvement than usual will be necessary while our new societal framework is being hammered out and even after. Improved ethical standards and enforcement will be important contributors to improved resolution. Certainly a major long-term, public educational effort is necessary. But the most vital contributions of all will undoubtedly come from the development of new approaches to dispute resolution. A number of legal scholars already believe that our judicial system simply is not equipped to handle these new kinds of problems. If they are right, critical institutional changes will be required.

In 1906, Roscoe Pound "urged that innovative methods of dispute resolution be devised." Over seven decades later, U.S. Chief Justice Warren E. Burger, "echoing Pound, insisted 'we should get away from the idea that a court is the only place in which to settle disputes.'"[27] Much has been done during the intervening seven decades, including the establishment of administrative and other specialized courts, the easing of wide public access to small-claims courts, the development of a great variety of specialized procedures for reaching accommodation between individuals and organizations, and more. But ordinarily, any really effective new approaches have been instituted only in response to catastrophe and emergency, such as the Great Depression, World War II, or Watergate. For a time, it appeared that the energy crisis might provoke such new approaches once again. But, for the moment at least, that problem has faded into the background, and we have returned to legal "business as usual."

We need a major ongoing effort to develop new ways to resolve socioscientific disputes—not only in response to emergency. Things may be quiet for the moment, but the critical problems will not go away.[28] Lawyers, law schools, bar associations, and all other segments of the legal profession must study, discuss, debate, write, and explore this area. Herbert Marks, a valued legal colleague, says, "Civilization means trying to do better." It is impossible to tell just where "doing better" will lead, but the same legal genius which can marshal millions of documents and thousands of pages of witness testimony into brilliant argument, can surely produce better ways than we have now.[29]

In 1973, Stanford Law School hosted a major interdisciplinary science, society, and law conference. The legal keynote began and ended as follows:

> One of the most exciting and rewarding roles the lawyer plays is that of architect of the structure within which society lives. With each new development in history, the lawyer has often been out in front, advising and guiding along new paths. He has at least a dream of leading society towards a better life. . . .
>
> But let us also recognize that we are embarking on a brave voyage in uncharted waters, and that we are being given the opportunity to help lead the world into an exciting and wonderful new age. Let us therefore also dream a little.[30]

Lawyers may be representatives and not principals, but that should not stop them from trying to influence their principals. Society is patiently awaiting their dreams and leadership.

4

CORPORATE RESPONSIBILITY

If there are to be voluntary, "self-serving" limitations on the use of adversarial extremes in socioscientific disputes, I believe that the initiative must come from the corporate party. None of the others involved has the necessary power or incentive. The attorney, as we have seen, is a representative or agent of his client, not a principal. He has no mandate or right from society to take action of this kind. The environmental, consumer, citizen, or "public-interest" participants are doing just fine. They lose as well as win some, of course, but overall their batting record is a fine one. The "government"—local, state, or federal—has no single position. It represents the particular political philosophy of its elected officials, and is necessarily subject to a variety of pressures of all kinds. They also provide just as much incentive for it to "win," as any of the other disputants.*

The large corporation is an almost inevitable participant in these disputes. *It* has the incentive to change, because it is doing so badly. *It* has, or should have, the broad, long-range understanding of its

* Socioscientific dispute resolution usually involves very long-term issues. It requires a similarly long-term underlying philosophy, which should not be subject to change with political elections. Moreover, as stated in the preface, I believe that we should turn to government only for solutions of last resort, where other reasonable alternatives are lacking. Many thoughtful persons, however, believe that government should take the *lead* role in societal affairs of this kind. I have not sought to argue the merits of my own political/social philosophy in this book.

total involvement in society. *It* can frame its position by balancing short-term consequences in particular limited disputes against consequences far on down the line—at other times, in other places, and in entirely different circumstances. *It* is the principal with the clear power to act, not the legal representative charged with the duty of doing what the client wants.[1]

Most important, in contrast to the lawyer, the corporation has a "public-interest" ethic which can tell it how to resolve sensitive issues with regard to such matters as the use of adversarial tactics in socioscientific disputes. That ethic can be found in the concept of "corporate responsibility."

"Corporate responsibility" is a presently much used, and much abused, phrase.* Its only generally accepted definition is that corporate action should be in the "public interest," but as we have seen in other connections, this kind of general definition poses more questions than it answers: How does one define "public interest?" Is it simply obeying the law and making a profit for the owners of the corporation? Or is it something more, calling for voluntary action with some social purpose other than obedience and profit? If the latter, how should the organization decide on what the social purpose should be?

For present purposes, corporate social responsibility can be analyzed in three behavior models. Those are the "activist," the "conservative," and the "business" models. It is the last of these—the "business model"—which can call for employment of the "rule of reason" in socioscientific dispute resolution, and which thereby introduces more sense into the process.

As with the earlier discussion of "risk/benefit," an historical perspective will be helpful to an understanding of "corporate responsibility." It has not always meant the same thing, and is quite clearly changing today in a dramatic fashion.

The corporate legal form has long historical roots. It burgeoned with industrialization, because of the efficiency with which it permits commerce to be conducted. That same efficiency has naturally led

* The treatment of "corporate responsibility" in this chapter is strictly in a dispute-resolution context, and does not consider the many difficult substantive political, social, and economic issues involved. I also use the terms "corporate" and "corporation" for convenience. However, the key analyses here and elsewhere in this book include other similar forms of delegated management, such as associations, trusts, and limited partnerships.

also to growth and increased capital resources; and with size and strength, came power. Corporations have proved to be no exception to the attacks on concentrations of power which have always been mounted by segments of society which consider themselves aggrieved. In the past, these segments have brought down primogeniture (the inheritance of property by the firstborn male) in England, the influence of the Church during the Middle Ages, and even monarchies and dictatorships.

In this country, the attacks (against the "robber barons," the oil "trusts," and others) began well back in the last century. The first results were specific limitations and controls on the powers of special groups, such as banks and railroads, considered to pose special problems. Significant general restrictions on corporate activity began with the Sherman Antitrust Act in 1890, following which came an increasing variety of new limitations, such as the Clayton Act, the Federal Trade Commission Act, the Robinson-Patman Act, the Securities and Securities Exchange acts, the National Labor Relations Act, and a great number of other more specialized statutes.

Laissez-faire, in the Adam Smith sense, was gone long before World War II; but compared with postwar restrictions, the early effects of all these controls were minor. The public-interest ingredient of corporate responsibility, as expressed in prewar limitations, was pretty much understood to mean only obedience to the specific requirements of law, plus profit-making. Indeed, the management of any large and publicly owned corporation which sought to employ substantial amounts of corporate funds to pursue some strictly social objective, or which refused to undertake a profitable venture because it disagreed with the policies of a particular foreign government, assumed a clear risk of suit for "waste" of corporate resources, or mismanagement.[2]

In other parts of the prewar world, the response to corporate power was often far more extreme than here. Some countries were simply unwilling to permit private corporate power to remain, whether subject to statutory limitations or not. Ownership and control were vested outright in the state. Fortunately, it has not yet happened here.

As with the public perception of the impact of science on society, World War II marked a turning point in this country as regards the role of the corporation in society. The same scientific and technological discoveries which led to the introduction of so many new developments also resulted in enormous and virtually unprecedented increases in the size and influence of commercial organizations. Terms such as "conglomerate," "congeneric," and "multinational" came into common legal and commercial parlance. Many of these organizations had more income, net as well as gross, than some foreign countries. They directly affected and influenced the lives of millions of employees, suppliers, and customers; and, indirectly, the world economic and social order generally.

These developments generated increased public concern in this country about the sometimes seemingly uncontrollable influence and discretion of the giant corporation. The demand for social responsibility in corporate action grew. This time, however, it meant something far more than simple obedience to the law, and profit-making.

The concept of corporate responsibility still has no very precise dictionary definition, other than that corporate conduct must be in the "public interest." Clearly that public interest requires obedience to the law and some measure of economic success, as in the prewar years. But many people using the term also mean that the corporation must "do what is right" in a much broader sense. The social "activist" commonly speaks in terms of some specific social or political goal or goals. The "conservative" or "traditionalist," in contrast, considers that profit maximization alone best serves the public interest, and that no other goal needs consideration. A growing number of academic and industry leaders is endorsing a third view, that the corporation's social goals should be determined in terms of "long-run" profit maximization. Although this view is similar to that of the conservative, in that it emphasizes profit-making, it also permits consideration as well of the social concerns of the activist by introducing the element of the "long run." Some further discussion of these three models of corporate-responsibility conduct is appropriate.

The "Activist" Model

If one shares the social goals expressed by the activist, the activist model of corporate responsibility is superficially appealing:

The appeal of the Vietnam War activist:

Dow sells napalm for use by American troops fighting the Vietnam War. Napalm is a horrible weapon. The Vietnam War is bad. Therefore, Dow must stop selling napalm to the government.

The protest of civil rights activists at my middle son's college graduation in May 1979:

Wesleyan University invests in the stocks of corporations which do business in South Africa. South Africa practices apartheid. Apartheid is bad. Business in South Africa helps the South African government survive. Therefore, Wesleyan must stop investing in stocks of corporations which do business in South Africa.[3]

Appealing as some of these goals may be, the model goes much too far. It vests powers in corporate management which are far beyond those which should be placed in strictly private hands. If a corporation is given the right to decide not to sell to its government simply because its management opposes the Vietnam War, doesn't it also have the right to refuse to support government action against Idi Amin in Uganda, Ayatollah Ruhollah Khomeini in Iran, or Adolph Hitler in Germany? Do we really want corporations to maintain the equivalent of their own private state and defense departments, deciding on whether to approve or disapprove foreign policy? Shouldn't decisions of these kinds, and other broad social and political judgments as well, be made by the duly selected representatives of the people, against whom the people can take action (as they have done, so effectively) to end a war, or to remove a President? These questions answer themselves.

One high federal court, speaking of just this kind of a matter, said:

> There is a clear and compelling distinction between management's legitimate need for freedom to apply its expertise in matters of day-to-day business judgment, and management's patently illegitimate claim of power to treat modern corporations with their vast resources as personal satrapies implementing personal political or moral predilictions.[4]

A law school colleague put it this way:

> In a nutshell, the question is whether we confuse our own thinking by attempting to apply to corporations and other collectives (governments,

labor unions, etc.) the standards and terms of moral conduct that were developed—largely in a religious context—to govern the behavior of individuals. . . . Large corporations (and other collectives) are formed to pursue specific objectives and are made up of many people whose interests and motives may be quite disparate. To expect these amorphous bodies to display "honesty", "generosity", or "social conscience" —as those terms are conventionally applied to individuals—is, in my view, to lose ourselves in a semantic swamp. The problem is complicated, too, by the fact that the responsible members of collectives (corporate officers, labor union leaders, public officials, even law school deans) typically think and act in a *representative* way that often reflects their perception of their constituents' interests rather than their own convictions or preferences. Thus a company president may think it entirely "moral" to maximize profits for the benefit of his stockholders and colleagues, even at the expense of the consumer or the environment.

All of this is pretty speculative, of course. But I do think we ought to have better ways of talking about groups that rely less heavily upon the vocabulary of individual thought and conduct.[5]

The "Conservative" Model

In contrast to the activist model, the conservative or "traditional" model goes too far the other way. It represents in effect an abdication of the corporation's responsibility to be an active participant in society. It asserts that social goals are for individuals and governments, not for a corporate management which is using other people's resources and not its own. The only social purpose of the corporation, it insists, is to make a profit. Professor Milton Friedman is often quoted to the effect that the social responsibility of business is "to use its resources and engage in activities designed to increase its profits so long as it . . . [does not engage in] deception or fraud."[6]

In today's atmosphere of growing social consciousness, this prewar approach is as simplistic as that of the activist. It ignores the fact that corporations have no alternative but to make decisions every day on whether or not to sell military supplies, or to do business in a foreign country, or to install pollution controls beyond those required by current law. Choices have consequences, which responsible management must evaluate if it is to do its job properly. If opening a plant in South Africa means that many universities, pension funds, and others holding stock in a corporation will immediately dump it, and that this might force the market price to a low level at which acquisitions are impossible, and at which personnel incentive plans are destroyed, a management which fails to think about these consequences acts at its peril. What it does is another thing, but failure to recognize that socially disapproved conduct will have damaging consequences is to close one's eyes to reality. It surely constitutes mismanagement in the modern commercial sense. It also can produce as negative effects for society as the conduct of the activist.

How can corporate social decisions take full account of the increasing social demands of society and still preserve the incentive motivation of capitalism? How can this be done without constituting the corporation the "personal satrapy" of management? The answer is by requiring that corporate decision-making in areas of social concern be rendered in terms of long-run profit maximization. This is the approach of the third and final model of corporate responsibility, the "business" model.

The "Business" Model

The business model of corporate responsibility finds the corporation's public-interest ethic in the principle of "long-run profit maximization." That principle calls for obedience to present law, plus profit-making, of course. It also includes, however, examining where society will be next year, the year after, and as far into the future as one can reasonably forecast. Building a multimillion-dollar plant in compliance with today's pollution laws alone is not good business if it can be anticipated that the requirements will get more stringent and some present accommodations are reasonable; or if the immediate community is so antipollution that local support will be lost, labor will be difficult to employ, and employee morale will be destroyed. Organizing an enterprise in a foreign country is not good business if it can be anticipated that a revolution will soon end all foreign ownership, or if world antipathy to the policies of that country is likely to result in a boycott of its exports.

These are not just "dollars and cents" decisions, for morality, ethics, patriotism, and similar societal attitudes must also be evaluated. Difficult as it may be, the public's probable response to corporate conduct must be predicted. At least part of that response will inevitably be based on whether the public believes that the organization has been a "good citizen," in addition to a successful one.

Corporate management's own special values and ideologies must of course be considered in long-run decision-making, because they are relevant as suggesting where the public interest may lie. Management must also examine the perceptions of others—conservatives, liberals, activists, environmentalists, consumerists, women, blacks, and all other segments of society. Will people boycott the company's products? Will the labor pool dry up? Will serious environmental or consumer or civil-rights opposition develop? Answering these and many other similar questions is certainly a difficult assignment, especially if the management group is made up primarily of "insiders," with no minority group or special-interest representation. It is no more difficult, however, than a number of the other hard decisions the management of great corporations must make. It can and should be an exciting and rewarding task, involving the manager and director in the forefront of social trends and opportunities.

There is important support for this approach within the academic community. A decade ago, then business school professor George Cabot Lodge observed:

> Business cannot sell to a sick society, and improving the general quality of life offers business new and rewarding opportunities. Businessmen are mindful of their consciences and their images. Also, the penalty for neglect is high in terms of public outcry and government reaction.[7]

Georgetown Law School Professor Donald E. Schwartz has taken this a step further. He adds that

> if dedication to the pursuit of higher earnings will impair the continued ability of private enterprise to enjoy acceptance, perhaps the interest [of shareholders] is best served by tempering the pursuit of profit.[8]

Courtney C. Brown, dean emeritus of the Columbia University Graduate School of Business, urges that

> business, by action and word, must convey an awareness of the societal as well as the material services it has provided and has the opportunity to provide on a broadening scale to the public.
>
> More importantly, business must articulate more clearly its opportunities and its commitment to respond to the public's aspirations.[9]

What about the business community? It is my personal belief, from observation and working with business leaders, that there is increasing business support for the business model. Many of my colleagues in business, as well as in academe and law, disagree—some vehemently. Even some business people who say they personally endorse the model, contend that the words of their own business colleagues quoted below are mere "PR" and "image." In their view, such words frequently are contradicted by a wink, a smile, or a subtle reward or punishment.

My experience supports my belief, but is certainly not sufficiently broad to support a conviction. Business acceptance of the long-run profit maximization model is an important step in improving socioscientific dispute resolution. If I am wrong that the following pages evidence a valid trend, I hope that they will serve at least to help persuade the business community that such a trend *should* be pursued.

The Business Roundtable is a group of 180 persons, each of whom is the chief executive of a major corporation, most of which are "Fortune 500" companies. These are the persons most likely to be involved as ultimate decision makers in the really major socioscientific disputes. In 1978, the Roundtable issued an important formal statement regarding business practices. The part dealing with corporate responsibility is worth quoting in some detail. That is because the statement is not just language and public relations, as suggested, but admissible court evidence as to the reasonable standard of conduct to be expected of major corporate management. It is persuasive not only as to the companies whose executives were involved in its preparation, but for big business generally. Under the heading "Corporate Social Responsibility," it states in pertinent part:

> Another major board responsibility is the consideration of significant social impacts of corporate activities and relatedly the consideration of views of substantial groups (other than share owners) significantly affected by such activity.
>
> The board's responsibility is to direct the enterprise in the interest of the owners, subject to the constraints imposed by law. However, the interest of share owners cannot be conceived solely in terms of short-range profit maximization. The owners have an interest in balancing short-range and long-term profitability, in considering the political and social viability of the enterprise over time and in adjusting to the global

> environment in which it operates. Moreover, share owners and directors alike have an interest in assuring that entities with which they are identified behave ethically and as good citizens. It is the board's duty to consider the overall impact of the activities of the corporation on (1) the society of which it is a part, and on (2) the interests and views of groups other than those immediately identified with the corporation. This obligation arises out of the responsibility to act primarily in the interest of the share owners—particularly their long-range interests.

The Roundtable statement continues by considering the methods by which such decisions are to be made, and then provides:

> Corporations, corporate boards and corporate share owners, are not the right bodies to resolve, on their own, for example, issues involving relations with other countries or U.S. military policy. . . . The political process by which law is enacted provides an opportunity—although an imperfect one—to weigh the costs and benefits of complying with new social demands. . . . [M]ajor environmental impacts, equal employment opportunity, important relationships with communities or governmental authorities, issues such as foreign trade or investment activities affecting this country's international relations, and matters of comparable magnitude are part of the operation of a business and proper for board consideration.[10]

The Roundtable statement reflects the stated views of many business leaders. William Sneath, chairman of Union Carbide Corporation, expressed it as follows in an address to the 1978 Southern Governors' Conference:

> Most corporate managers regard social responsibility as a mandate. We know that our long-standing and legitimate rights to pursue private business objectives are subject to erosion if those objectives—and our practices—are not perceived to be in the public interest. . . .
>
> The public sector has invaded the private as never before in our history, a phenomenon that has made knowledge of and involvement in the public policy process an inescapable part of a business manager's responsibility.[11]

Fred T. Allen, chairman and president of Pitney-Bowes, Inc., included the following remarks in an address to the Swiss-American Chamber of Commerce in Zurich:

> As businessmen, we must learn to weigh short-term interests against long-term possibilities. We must learn to sacrifice what is immediate,

> what is expedient if the moral price is too high. What we stand to gain is precious little compared to what we can ultimately lose.[12]

J. Irwin Miller, chairman of the Executive Committee of the Cummins Engine Company, has said in encouraging corporate philanthropy:

> We save ourselves, our business, only by making this society work equally well for all its members. To me that means, among many other things, voluntary individual and collective concern, voluntary giving—giving knowledge, time, money wherever we are convinced it will improve quality, correct evils, extend equity in America. The case for corporate giving is an essential part of corporate survival. . . .
>
> I think such an America might be a very good place for business. So money and time and concern spent [on the arts and on helping the less fortunate] might turn out really not to have been "spent" at all.[13]

And Pillsbury's Jerry Moore concludes that social responsiveness "is no longer a matter of corporate (company) responsibility, it is becoming a matter of corporate (industry) necessity."[14]

A much respected business associate points out that long range profit maximization is perfectly consistent with the industry aphorism "let the marketplace decide." Those automobile manufacturers who were best able to anticipate environmentalism and government regulation, the Israeli-Arab conflict, the Arab oil boycott, OPEC and the energy shortage certainly won critical market acclaim; those who failed to include these long-range general societal developments in their planning, including once world-renowned American giants, lost vital market shares.

A number of persons have urged that "corporate responsibility" be given enforcement teeth. W. Michael Blumenthal, while chief executive of the Bendix Corporation, called for the organization of a new "institute or association to promote the idea of responsibility and ethics in business practices in the broadest sense." This group "would focus on devising new ethical-behavior codes to which all business would be expected to subscribe."[15] Arthur Schlesinger, Jr. (Albeit Schweitzer Professor of the Humanities at the City University of New York) and Irving Kristol (Henry Luce Professor of Urban Values at New York University) both endorsed this approach. They have suggested that those who violate corporate ethics be "read out" of the corporate community.[16] These proposals pose antitrust concerns, and may not be practically possible at present. However, they do reflect a developing conviction that the time for meaningful corporate social responsibility has come.

As with the Business Roundtable statement, words of this kind from senior business executives are important in themselves. Despite their perhaps necessary generality, they do represent "operative" instructions to corporate subordinates. Although it is too early for the evidence to be sufficient to render final judgments, there are indications that the instructions are being followed. At the end of 1979's first financial quarter, *Business Week* reported that about half of the 120 large corporations it had surveyed discussed corporate social reponsibility and key political and environmental issues in their reports.[17] Allied Chemical Corporation has been putting its managers through a three-day seminar, in which they consider the obligations of the company to society at large.[18] More and more companies have "public interest" and "social action" committees, and other groups with delegated authority in the public-interest area. In the early 1970s, the Dow Chemical Company adopted a voluntary antipollution campaign, setting up systems for recycling chemical waste at several of its plant locations, at no net cost. Although I suspect it might not be able to duplicate the results in today's far more pervasive regulatory environment, for a time it even managed to realize a profit on the venture—thereby providing objective evidence of "long-run profit maximization."[19]

One overriding criticism of this manuscript has been that corporate management simply will not act in accordance with the business model—that short-run, day-to-day profit maximization is, and will continue to be, the guiding principle for all corporate conduct. Even some of my "Wall Street lawyer" colleagues, with substantial corporate practices, believe this to be the case. If only those industry leaders who share that view and act in that way would take the time to listen to their own children condemning such conduct. If only they could hear the disapproval of my law students—among the cream of American society, who will be running this country a quarter-century hence. Ignoring such attitudes is to say, "Après moi, le déluge!"

I am not suggesting that corporate responsibility, in terms of long-run profit maximization, is yet the guiding spirit of American big business generally. Without doubt, we have a very long way to go.

Those who are recalcitrant or have doubts will need convincing. But the concept does appear to have solid support. It is gaining momentum. It is a solid foundation upon which to build a new approach to the resolution of socioscientific disputes. How to do this is the subject of the next chapter.

RESPONSIBILITY IN DISPUTE RESOLUTION

Adversarial extremes in socioscientific disputes are dangerous. Any corporate management applying "long-term profit maximization" principles, would clearly prohibit them. If short-run profit maximization is applied, the case for total prohibition may be somewhat less clear. The use of extremes does backfire every so often, however, with disastrous consequences. Even here, then, sound corporate management would insist upon adequate control over the employment of such extremes, in order to be sure that the risks of use do not outweigh the benefits.

Lawyers and other dispute-resolution representatives cannot be expected voluntarily to abandon these tactics, at least for the present. It is accordingly the task of senior corporate management to insure dispute-resolution compliance with its corporate responsibility policy. That is a difficult assignment. Doing the job will call for changes in traditions with a long history. It will take a long time. At the inception, it will involve senior officers, who ordinarily attend only to policy matters and leave tactics to others. They may be required to participate in some aspects of corporate operations. It will mandate the introduction of tight management practices into the administration of dispute resolution, including reporting, auditing, and investigating procedures. There must be changes in the relationships between the corporate principal and the dispute-resolution representative, primarily the attorney, so that there can be no doubt where authority and responsibility lie. This chapter deals with those subjects.

Before turning to a discussion of specific practices and proposed modifications or changes, some consideration of the overall management framework will be helpful. It will explain why, in this case, it is so difficult to translate management policy into organizational practice.

Resistance to Change

Applying corporate responsibility to dispute resolution is, in practice, difficult because it is so clearly contrary both to the general pattern of organizational rewards and punishment, and to our historical tradition of independence in dispute-resolution representation.

Presidents and chairmen can talk about corporate philosophy, but others on down the line are frequently rewarded on the basis of more traditional performance. They are promoted, or receive salary increases in terms of how well they met specific "dollars and cents" targets during the last fiscal quarter, in sales, profits, or costs; or how they performed in terms of market shares, personnel turnover, new product development, license approvals, and the like. Performance incentives of this kind may lead even to "technical" violations of law, and sometimes worse. I have already referred to the government criminal prosecutor, who admittedly ignored the possible requirements of Swiss law in order to obtain a conviction he believed was sought by his superiors. Before passage of the Foreign Corrupt Practices Act (FCPA), the incentive to "please" his superiors and to "succeed" was the same for the corporate manager in a foreign country. A questionable payment to a foreign official sometimes appeared to be the only way in which he could market his company's product. He believed that this alone would enable him to achieve the goals his company had set for him. He was not at all happy at the prospect of paying a bribe.[1] Yet at the same time, it was a "way of life" and "everybody is doing it."[2] It is only on the basis of hindsight, or with a great deal of grandstand quarterbacking by those removed from the competitive struggle, that we have finally today begun to enforce measures designed to prohibit certain of these unlawful foreign practices.

The corporate enforcement problem is far more difficult when the questioned conduct is not unlawful by any test, and is only prohibited by the philosophical concept of corporate responsibility. The corporate manager may be presented with an obviously profitable opportunity in South Africa, or Uganda, or Chile, or Cuba, or Vietnam, or Iran, or someplace else. Discovering the possible long-term "negatives" to the transaction may require an in-depth investigation of many remote subjects, perhaps including American foreign policy. He will be understandably reluctant even to conduct the investigation, much less to reject the proposal or to send it along to senior management with a full investigative report for evaluation.

If corporate management wants to be sure that its representatives obey its "social responsibility" rules, it must act positively and dramatically. It must make clear that violations of these rules are no more to be tolerated than the unlawful political contributions and foreign bribery which were once countenanced.

The beginnings of management control over the implementation of social-responsibility directives are starting to appear. More seem clearly on the horizon. Many companies already employ success/failure tests in activities which are not strictly for profit, such as with respect to product complaints, favorable or unfavorable publicity, worker-accident experience and minority-employment rates. Even broader proposals have been made for the equivalent of a "social audit," which would provide objective measures for progress in reaching other social objectives.[3] But these are all still quite rudimentary. Until they are developed and refined, any senior management which means business must maintain an intimate involvement in this area of operations.

When the social goal being sought is in the dispute-resolution area, strong contrary traditions of attorney independence must be added to this incentive opposition. There is a growing number of exceptions, but for the most part it has always been common practice for clients to turn even major matters over to counsel, lock, stock, and barrel, with but the most limited of instructions as to how to perform. Many consider it somehow "unprofessional" to try to tell a lawyer how to do his job. Many lawyers resist client participation, much less instruction.[4]

The practice is much the same within wholly internal corporate law departments themselves. Such departments are growing rapidly today.[5] However, they are still relatively new. Most senior managements tend to accord them great freedom and discretion. There is as yet little history of tight management even of this internal legal function. There are very few books instructing in this important area[6] and there are fewer business schools offering any training at all in it.[7] Outside law firms retained to assist in socioscientific dispute resolution are even less tethered, either by the corporate law department, or by senior nonlegal management.

The lay person to corporate management, or to the law, may assume that the internal corporate general counsel is always a member of senior management and is intimately informed about all its policies and directions. That is sometimes the case, of course. Quite commonly, however, the general counsel is not a member of the really senior management team and may not know whether, or how, his senior management wishes some general policy expression implemented. Surprisingly, perhaps, sometimes the general counsel does not even have the authority effectively to control outside counsel if he wants to. Outside counsel may have been selected by senior management, or sit on the board of directors, with a much better pipeline to the top.

The almost inevitable consequence of all this is that senior management frequently knows very little about even the most seriously damaging conduct taking place in socioscientific disputes involving their companies. In Allied Chemical Company's Kepone debacle, *Fortune* reported on the "smoking gun" (a memorandum by one Virgil Hundtofte, an Allied plant manager) which so damaged Allied, as follows:

> It seems clear that Allied's top management was not aware of Hundtofte's decision and would probably have rejected the scheme had they known of it. The Justice Department, which had access to all of Allied's files relating to Kepone, was unable to turn up any convincing evidence that corporate management had either encouraged or approved of this evasion. . . . One lesson that all companies may draw from Allied's troubles is the need for tighter control at the top to prevent the

> type of massive litigation that can burst from seemingly routine decisions at the lower level.[8]

In its litigation with Berkey Photo Inc., it appears that Eastman Kodak had no knowledge of the unlawful tactics in which its outside counsel was engaging. This included suppression of evidence and even outright lying, for which a senior partner in the outside law firm handling the case went to prison.[9]

In the General Motors/Ralph Nader controversy, James M. Roche, then president of General Motors, testified in Congress to his lack of knowledge about what was going on, as follows:

> Let me make clear at the outset that I deplore the kind of harassment to which Mr. Nader has apparently been subjected. I am just as shocked and outraged by some of the incidents which Mr. Nader has reported as the members of this subcommittee.
>
> As President of General Motors, I hold myself fully responsible for any action authorized or initiated by any officer of the corporation which may have had any bearing on the incidents related to our investigation of Mr. Nader. *I did not know of the investigation when it was initiated and I did not approve it.* (italics added)[10]

Whether known to senior General Motors executives or not, it is the public perception that General Motors continues to employ similar adversarial legal tactics today, over a decade later. One 1978 news article, summarizing a number of cases attacking GM's credibility, concluded:

> G.M.'s strategy in court seems to question the manner in which the cases have been investigated rather than the merits of the allegations themselves. . . .
>
> Based on its current legal strategy, then, General Motors may never have to meet the issue of integrity head-on.[11]

There is not a great deal of additional positive evidence on which to base the conclusion that senior management is often uninformed regarding important dispute-resolution tactics. Executives, as well as attorneys, understandably treat this kind of information as sensitive and confidential. It is not surprising that those who do feel compelled to talk about such matters, almost uniformly deny knowledge of, and participation in, conduct which is charged to be improper and even unlawful. For some years, however, I have been teaching a law-school seminar on socioscientific dispute resolution. Each of the students conducts intensive research into a related subject, usually a specific dispute. In little of their work have I seen any significant evidence that the senior management of other large corporations knew much more about what was going on in socioscientific disputes than did the executives of Allied Chemical, Kodak, or General Motors.

If socioscientific dispute-resolution tactics are to conform to corporate responsibility principles, we must institute changes which will insure that senior management is made to know what is going on, so that it can take steps to see that any contrary corporate responsibility policies are enforced.[12] The excuse, "I didn't know," cannot be accepted except in these few cases in which the dispute-resolution representative is on such a "frolic and detour" of his own, that responsible management controls fail.[13] It may be that, in some cases, senior management would like to engage in "image building," and not want its expressed social policies enforced. An "upwards" information policy will help prevent this. That is because any contradiction of stated policy by the wink, or the smile, or the pat on the back, will be made obvious to management by the clear disparity between policy and practice. The policy will thus place corporate responsibility right where it should be, at the top.

Dispute-Resolution Authority

Until long-standing tradition and present practice have been changed, tight management from top to bottom is the only way in which to insure effective application of corporate social responsibility principles in socioscientific disputes. Key senior management must hold the final reins. The common practice of delegating virtually unlimited authority to dispute-resolution representatives—inside or out, attorneys or not—simply will not work.

As in so many other respects, the IBM dispute with the federal government furnishes the outstanding example of both problem and solution. For years, IBM has been attacked for dispute-resolution tactics which to many appeared to be obstructive, dilatory, and worse. Outside counsel, inside counsel, and the company itself were all targets.[14] It got so bad that the front pages of the trade press, and even the general press, reported charges of improper delaying tactics levied against IBM by the Assistant Attorney General in charge of the Justice Department's Antitrust Division, and the issue of delay was aired before a congressional committee.[15] One could only guess at whether IBM's outside counsel, inside counsel, senior management, or any combination of the three, was responsible, but the damage to IBM's credibility and reputation was the same in any event. If top management had specifically authorized the tactics, it undoubtedly had not been made to appreciate their consequences.

Finally, senior IBM management apparently decided that enough was enough. At IBM's annual meeting in Denver in late April 1978, IBM Chairman Frank T. Cary devoted half of his formal address to the case. He told the stockholders: "I want you to know that I have instructed our lawyers to present IBM's case as crisply and expeditiously as possible."[16] There had been talk of this kind before, with IBM charging that delay was the fault of the government, not of itself. This time, however, the talk was followed by action. When the Government finished its presentation of evidence in the case shortly thereafter, *Electronic News*, a leading weekly trade newspaper, reported with regard to implementation of the Cary instructions:

> To that apparent end [IBM trial counsel] surprised many of the approximately 30 persons in the courtroom last week by not filing a motion to dismiss the government's case. He told the judge that, "even though we believe there would be merit to such a motion," he would not file it because of the long delay it would create.[17]

IBM also pared down its witness list, in apparent further evidence-tightening moves.

In mid-1979, IBM concluded that it should make application to disqualify the federal judge who had been presiding over the case. It had concluded that he was so prejudiced against IBM that a fair trial was no longer possible. Although I am not privy to IBM's intimate planning, clearly it must have recognized that such an application would be attacked as just another delaying tactic, and that is what happened. But, this time, IBM made very clear that the move was a product of a clear corporate board of directors "public-interest" decision. It was not simply another adversarial legal tactic, designed to "win," without regard to the merits. It assigned the problem to five "outside" directors, all distinguished persons whose integrity and credibility could not seriously be open to question. All were attorneys, but not serving as IBM trial counsel in the matter. They included two former Cabinet ministers, Carla Anderson Hills and William T. Coleman, Jr.; a former governor of Pennsylvania, William W. Scranton; the chairman of E. I. du Pont de Nemours & Company, Irving S. Shapiro; and John N. Irwin II, former U.S. ambassador to France.[18]

By its dramatic action, IBM demonstrated just how senior management can enforce social policy in dispute resolution—simply by taking control. That is what its chairman did, and that is what its special outside directors' committee did. Whether or not one quarrels with IBM's proposed action (and, of course, there continued to be charges of delay and obstruction),[19] there no longer was any question as to who was responsible for what was happening, and who was in charge. There was a second important, albeit perhaps incidental, benefit to this demonstration of management control. At least with regard to the application to disqualify the judge, IBM communicated effectively that it was acting in what it genuinely believed to be the true public interest, and was not "game playing." IBM was unsuccessful in its move to oust the judge, but its public credibility did not suffer. Indeed, many of its most bitter opponents reluctantly conceded that there was merit to its contention. This contrast with the earlier history of the dispute, characterized by widespread dismay at IBM litigation tactics, should have been satisfying to IBM. It certainly encouraged the exploration of settlement, which had theretofore seemed impossible.[20]

IBM's management control of the reins is illustrative and important, in part because it stands virtually alone. The contrary practice, I fear, is the norm. IBM's own earlier history of apparent inadequate management control is persuasive evidence in this connection. So is industry's Business Roundtable's experience in trying to speed up decision-making in disputes in the energy area.

For years, industry has been complaining that unreasonable obstacles placed in its path by government and environmental and consumer opponents have caused delays in energy-related administrative proceedings. The result has been cost escalations which are unacceptable. In March 1978, the Business Roundtable brought all this to the attention of the President of the United States. In a very carefully prepared presentation, it described eleven specific energy-related projects which had been thus blocked, and urged adoption of a number of emergency remedial actions by government.[21]

Nowhere in the presentation was there any suggestion that any of the delay was caused by industry's own representatives, or any pledge by industry itself to take action to mitigate the problem. Almost certainly, the executives involved simply did not know that in at least some of the cases, industry's own tactics were culpable as well. Yet in one of the projects cited by the Roundtable, a utility itself provoked the opposition which resulted in loss. It sought to justify its refusal to respond to a leading environmental organization's repeated written requests for assurances, on the ground that it had no legal obligation, or "responsibility," to do so. Technically perhaps this was correct, but subsequent discussion of the utility's tactics revealed that they had aroused the suspicion, antagonism, and ultimately successful opposition of the environmental inquirer. The utility's own legal tactics may thus, in fact, have caused the disaster. As one environmental leader commented, "We are sustained by the arrogance of our adversaries." (See Appendix A.) In another of the Roundtable projects, "sporting" dispute-resolution tactics were equally a contributor to cost and delay.[22]

Dispute-resolution tactics do not present easy problems. Management cannot simply dispose of them by saying, "Don't delay," and "Don't obstruct," and be done with it. The very act of opposition obstructs the other side and necessitates delay in order to develop and present contentions, evidence, and arguments. To say and do nothing, and simply press the opposition to hurry up and complete its presentation, is not an acceptable alternative. Acquiescence is better.

As earlier stated, frequently both the purpose and the effect of dispute-resolution conduct are unclear. Production by an opponent of millions of documents, or disclosure during dozens of depositions, inevitably means delay—but it also may provide information which is critical to resolution of the dispute. Often the perception of the dispute-resolution representative who is intimately involved in the battle is suspect. He wants so desparately to win that he may not be able adequately to take into account other concerns. Management must get into the matter with both hands and both feet, or at least be certain that all close questions are referred back on up the ladder of authority until someone is able to provide a clear and acceptable answer.

Where a socioscientific dispute is in the hands of attorneys, as is commonly the case, the objection is sometimes made that lay persons are not competent to make decisions concerning document production, depositions, and other such strictly legal matters. That is nonsense. Corporate management makes decisions in the most complex of areas every day, without apparent trouble. It requires only that the expert lay out all the options and all the consequences which he can anticipate. If anything, it is the dispute-resolution attorney who may not be competent to judge long-run negative effects of dispute-resolution conduct. There may be consequences of which he is not informed, and which he is thus not competent to evaluate, but which can be most damning at some other time and place, or in different circumstances.

The Rule of Reason

Thus far in this chapter, I have discussed management control of dispute-resolution tactics, irrespective of management's corporate responsibility policies. As stated at the outset, tight control is necessary whether short-run or long-run profit maximization principles are to be applied.

If management policy is long-run profit maximization, however, management involvement in dispute resolution can be much simplified. That is because, under such a policy, the use of adversarial extremes would be totally prohibited. Then a specific tactical dispute-resolution philosophy, the "rule of reason" referred to in chapter 2, can be applied. Once this is adopted by management and communicated on down, any competent dispute-resolution representative will know how to act. Senior management need not thereafter be concerned with the intimate details of dispute-resolution tactics.

The fundamental precept of the rule of reason, is consistency between all dispute-resolution conduct and all other corporate action, including expressions of policy. The rule has nineteen conduct guidelines designed to achieve this objective. Its central theme is that dispute-resolution tactics must be as much in the public interest as the outcome being sought. It thus translates the general concept of "long-run profit maximization" into real-life dispute-resolution practice.

Consider the unfortunate, but quite typical, dispute-resolution tactics for the defense of antitrust charges referred to in chapter 3. As recommended by the former senior government antitrust attorney, now in private corporate law practice, they would include the filing of a "counterclaim" and the taking of "exhaustive discovery" in order to bleed the other parties and compel them to "look at their pocketbooks rather than the so-called 'public interest.'" It would also include deliberate delay to force the opposition legal staff to "turn over a couple of times." Obviously there is at least some possibility that these tactics might be successful, or they would not be so common. Nor would they be recommended by a distinguished trial lawyer to a sophisticated legal audience. Any corporate trial lawyer—especially one who is not informed regarding his client's long-range societal goals and objectives, or its other operations and problems—might well consider that his general assignment to "win" included the authority to employ such tactics. He might not even bother to seek permission.

Long-run profit maximization, however, does not permit such mechanical, or automatic, adoption of dispute-resolution strategy. Senior corporate management must retain control, so that it is informed of all aspects of the dispute-resolution process, tactical as well as substantive, in which important long-run policy concerns are involved. With respect to the present hypothetical antitrust dispute, it would have to consider whether the lengthy effort, the resulting uncertainty—which is sometimes more difficult to deal with than a loss—the enormous out-of-pocket litigation costs,[23] the diversion of personnel from other productive assignments, and the real dangers to corporate society generally of employing such tactics, are worth the possible "win." As discussed earlier, this kind of conduct can sometimes "blow up," with disastrous consequences. Management must ask, "Can we expect to win by playing it completely straight? If so, is there any real point to making the other side suffer? If not, can we get rid of the dispute by quick settlement, at a cost which includes something for getting rid of the risks? If maybe, let's do a comprehensive 'risk/benefit analysis,' which lays out all the pluses and all the minuses, and then make our decision."

One of the critical concerns which must be factored into any adequate long-run dispute risk/benefit analysis, is the effect of such "game" tactics on the credibility of the particular corporation concerned, and on corporate society generally. It bears repetition that the public regards adversarial extremes in socioscientific dispute resolution as antisocial. The result of their use has not only been a loss of credibility for particular corporations and industry at large, but increasing popular support for limitations on corporate discretion, freedom, and power. The regulatory controls of today, against which so many industry leaders inveigh, are but a taste of what may be if the demands for management by "public" and "outside" directors, for federal chartering, and for more extreme measures, find additional support. This might not be the state ownership imposed in some other parts of the world, but the effects would be similar.[24]

Surely industry must acknowledge that enlightened self-regulation is preferable to regulatory coercion. Underneath it all, that is a theme which pervades the principle of long-run profit maximization in the first place. In an analysis of sociologist Morris Janowitz's essay "The Last Half-Century: Societal Change and Politics in America," Naomi Bliven has written:

> The author's chief preoccupation is with social control—the ability of groups in society to regulate themselves both for their own benefit and for the well-being of the nation. Social control is a substitute for coercion, and it exists to some degree in all societies. A high degree of social control goes with a high degree of liberty. Janowitz, who wishes for a minimum of coercion and a maximum of liberty, feels that our "ordered social segments" have failed at the tasks of social control: although they may regulate themselves, they are disconnected from each other and from the nation. They do not or cannot accept sufficiently broad responsibility for solving the problems that clog the political process.[25]

The rule of reason bypasses this lengthy management analysis of dispute-resolution tactics. It is the specific application of the corporate-responsibility policy of long-run profit maximization to socioscientific dispute-resolution tactics. It measures all proposed conduct by the test of how it would appear on "center stage"—as though whatever was being said and done at even the most secret or intimate moments were on nationwide television, or on the front page of the local press, subject to view and re-view by even the most antagonistic adversary. It mandates that adversarial action must be consistent with adversarial position. There may be marginal decisions that are difficult to make, but ordinarily the question of how to act will be answered easily by the substitute question, "How will what I am thinking, saying, or doing privately here sound if it is published on the front page of the *Washington Post*, or presented on the "Today" show?" A counterclaim and exhaustive discovery to "bleed" an adversary and make it look at its pocketbook, or delay until the opposition legal staff has "turned over a couple of times"—however common such tactics admittedly may be—will not "go" by this test. By such a measure there would have been no Watergate.

This is not the book to discuss the details of litigation tactics, but a statement of the rule of reason's specific procedural guidelines will be useful—with the caveat, however, that there is far more to the approach than any brief summary will suggest. Despite what some readers might conclude from their overall tenor, for example, the following guidelines do not require the waiver of attorney-client privilege or trade-secret exemption, nor call for a party to "bare its soul" and "let the chips fall where they may." They do not necessarily preclude cross-examination of a witness regarding collateral matters, including even alcoholism or discrimination. The rule of reason is an adversarial dispute-resolution philosophy which is even more effective for the party employing it than the adversarial extremes it replaces. That is because it eliminates the inconsistencies resulting from tactical conduct, which impeach a party on the merits. The rule-of-reason guidelines are:

—Data will not be withheld because "negative" or "unhelpful."

—Concealment will not be practiced for concealment's sake.

—Delay will not be employed as a tactic to avoid an undesired result.

—Unfair "tricks" designed to mislead will not be employed to win a struggle.

—Borderline ethical disingenuity will not be practiced.

—Motivation of adversaries will not unnecessarily or lightly be impugned.

—An opponent's personal habits and characteristics will not be questioned unless relevant.

—Wherever possible, opportunity will be left for an opponent's orderly retreat and "exit with honor."

—Extremism may be countered forcefully and with emotionalism where justified, but will not be fought or matched with extremism.

—Dogmatism will be avoided.

—Complex concepts will be simplified as much as possible so as to achieve maximum communication and lay understanding.

—Effort will be made to identify and isolate subjective considerations involved in reaching a technical conclusion.

—Relevant data will be disclosed when ready for analysis and peer review—even to an extremist opposition and without legal obligation.

—Socially desirable professional disclosure will not be postponed for tactical advantage.

—Hypothesis, uncertainty, and inadequate knowledge will be stated affirmatively—not conceded only reluctantly or under pressure.

—Unjustified assumption and off-the-cuff comment will be avoided.

—Interest in an outcome, relationship to a proponent, and bias, prejudice, and proclivity of any kind will be disclosed voluntarily and as a matter of course.

—Research and investigation will be conducted appropriate to the problem involved. Although the precise extent of that effort will vary with the nature of the issues, it will be consistent with stated overall responsibility to solution of the problem.

—Integrity will always be given first priority.[26]

In a 1978 address to the Conference Board, Earle B. Barnes, chairman of the Dow Chemical Company, called for application of the rule of reason in this way:

In the public's eyes, a corporation's intent is purely fictional. The public's view of a corporation's intent is determined by three things:

What is said; what is written; and, most important, what is done. . . . [T]here are, in substance, two ways to litigate:

—The first—and the right way—is by sound data and a reasonable approach to show that what one is doing is right.

—The second way is to bring the law to the lowest level of conduct by attacking the other person in the most negative way possible—applying the questionable art of smear and intimidation. . . . We must take the positive approach, apply the rule of reason, examining at all times what we say, what we write—and giving even greater scrutiny to what we do.

If what we do is right, the truth will speak for itself and we will be judged by our actions. Rancor diverts attention; reason focuses attention on constructive acts. . . .

Truly, the rule of reason—properly exercised—can work communication miracles.

American businessmen and women have become global experts in selling their products. But many would have a very low Nielsen rating on their performance in selling themselves.

A few years ago, we conducted an employee survey in which we asked if our people felt they were well informed on company matters. One of the most interesting comments played back repeatedly was, "Give us the bad news as well as the good."

Does this not give us a basic clue to credibility? People are not afraid to face the facts when reasonably presented. Candor brings acceptance.

Let's apply the rule of reason. It is an important tool to help turn the regulatory tide and enhance the image of American business in a short time.[27]

Jack St. Clair, then president of Shell Chemical Company, has urged adoption of the same principles:

The only way we can build that credibility, I believe, is if, as time goes by, we conduct ourselves in an honest and open manner. . . . Internally, we must understand that our operations must be conducted in a totally honest and open manner, in accordance with not only legal but high ethical standards.

We must demonstrate by our conduct to the public as well as to our business associates, the high standards under which we ourselves operate. . . . I think high ethical conduct will be rewarded by public support for what we are trying to do.[28]

In early April of 1977, the National Academy of Sciences conducted a public forum. One of the workshops dealt with socioscientific dispute-resolution methodology, and the rule-of-reason approach. The unanimous conclusion of the workshop is relevant to many other aspects of this book, and the full report is reproduced in Appendix A. A part deserves repetition here:

> The so-called "establishment" must accept the primary leadership role in reducing confrontation and introducing reason and logic into the resolution process. Presently, there is extremism at many points on the opinion spectrum. Whether or not extremism is ever justified—and remember, it was just last year that England as well as the United States celebrated the bicentennial of our own revolution—it may be at least understandable on the part of the deprived and disenfranchised when they have no other means of getting attention. The thrust of the experiences related by [two industry executives] is that the leaders of science, industry, and society must respond to the extremism of "outsiders" with data, reason, and understanding. There can be no excuse for the all-too-common response of equal or even greater extremism—such as [examples given]. . . . All of this has contributed to the mistrust of industry, and science, and the establishment by a major segment of our society to the point where a good part of the public simply will not believe the unanimous assertions of the oil companies and industry regarding even this energy problem.

In the discussion of corporate responsibility, I found it unfortunately necessary to have to refer to the fact that many of my colleagues do not believe that senior corporate management means the fine things it says. The same is equally true here. My critics include a surprisingly large number of business lawyers with large corporate practices, and several law professors who are ordinarily quite favorably disposed to free-market principles and industry generally. One dear friend and valued colleague noted in his review of an early draft of this manuscript, "The trouble is that top management is long on high-sounding policies, but short on enforcement." The strong contention of these critics is that executives will mouth the right words to achieve some desired corporate "image," but when it comes to action, they want results—and good ones. They want their subordinates to yield no point, and give no quarter, and, where necessary, march right up to the border of illegality and even cross it sometimes—but not to talk about it. Their subordinates all know that they do not mean what they say. Everyone knows exactly what they do mean.

As with the discussion of long-run profit maximization, the available information about the professional behavior of top corporate executives is very limited.[29] Time alone will tell whether these warnings are valid. American industry has its culprits, just as do we lawyers and every other segment of society. Once again, however, my experience is that, on the whole, the key senior managements of our giant corporations—those which are the most frequent parties to the major socioscientific disputes—are sincere and dedicated professionals, who do mean exactly what they say, and have every intention of carrying out their promises. They need guidance and help in specialized areas, of course, but I would prefer to give them a chance to straighten out the mess in socioscientific dispute resolution. It is only if they fail that we must consider turning to such unfortunate alternatives as new regulatory or judicial interference.

The Lawyer-Director

If it is to achieve more satisfactory results in the resolution of socioscientific disputes, senior corporate management must assume control over dispute-resolution tactics as well as substance. This is so whether it adopts the "conservative" corporate responsibility model, or the Business Roundtable long-run profit maximization model. (There is no significant business support for the "activist" model.) In the former, it must participate much more intimately in the ongoing details of tactical decisions; in the latter, it can avoid much of that burden by simply mandating adherence to the rule of reason. In either case, however, it will, in addition, have to take steps to insure that it receives the information it needs, and that its directions are being followed.

As stated at the outset of this chapter, all of this will require changes in the usual relationship between a corporation and its dispute-resolution representatives, primarily lawyers. One important matter for consideration is the role of the dispute-resolution attorney in the corporation.

The senior attorney who is in change of the resolution of a socioscientific dispute in which his corporate client is involved must be as fully informed of the client's affairs as possible, including all of its long-run policies. Only in this way can he be sure that he will recognize problems requiring consultation and authority, or make "public-policy" judgments on his own where the course is clear. At the same time, however, he should act as a policy maker for the client in a socioscientific dispute most reluctantly and, if he does so, only with the greatest caution. He should sit with the board of directors when it meets, and know what it knows, for example, but, except under the most unusual circumstances, he should not also act as a director regarding such a matter.

There is much debate today in legal, government, consumer, and academic circles as to whether the attorney for a corporation should ever serve as a director as well. The view in favor of such dual service contends that he has critically important insights to contribute to overall corporate management, especially in these days of burgeoning responsibility to various publics. In addition, he needs all the information he can get to perform his legal assignment in the most effective fashion. He should have a right to information and to participation, as well as a duty and responsibility.[30] The contrary view is that there is an inherent inconsistency between the role of a lawyer and that of a director. The lawyer assesses legal risks and furnishes data and information to the director. The director is the risk taker. He decides what risks to assume. The lawyer may *advise*, for example, that a given course of action has a 40 percent chance of achieving a certain result. The director may *decide* that the 40 percent risk is worth taking, because the reward is so great. Those who oppose dual service, contend that it is simply impossible for one individual to "change hats" when dealing with a single matter. Matters are rarely as simple as my "40 percent risk" example. The lawyer must be free to argue in favor of his legal view. When he does, he inevitably becomes an "advocate." That precludes the kind of independent risk-balancing judgment which is required of him as a director.[31]

I do not think it is yet clear how this debate should finally be decided. Until we have more evidence, it is my view that corporations should be permitted to decide for themselves whether they want their counsel to serve also as directors. There are especially persuasive considerations favoring dual service for closely held ("private") or small "public" corporations. The balance presently seems to be the other way for larger public corporations. That is especially so where the corporate counsel sits with the board in any event and is given full access to all its information.[32]

Where corporations are involved in socioscientific disputes, however, dual service should be approached with special caution. Until there are major changes in approach and attitude, the socioscientific dispute-resolution attorney's goal is to "win" the specific battle at hand. He is paid on that basis. His reputation is made on that basis. Whether he is an "inside" or an "outside" lawyer, his colleagues, associates, partners, or other clients will not understand if he yields some important point to achieve some evasive "corporate responsibility" or other social objective. All of his training and instincts are opposed. I am not suggesting any legally enforceable preclusion of dual capacity. Where socioscientific disputes are involved, I am suggesting a management policy which permits dual service only in exceptional circumstances, because an attorney can not be expected always to make sound and impartial long-run societal decisions as a director where his contrary concerns are so strong. I hope that one day our dispute-resolution tradition will have sufficiently changed so that this need not be the case.

Identifying the Client

A corporation is a legal creation. It is made up of employees of all kinds, including day laborers as well as senior executives and directors. Many others have interests in it, including shareholders, creditors, customers, suppliers, and even members of the public who may be affected by its production, distribution, and other activities. Another important management function today must be to identify the specific interest which the dispute-resolution representative represents. Just who is the corporate client? Answering that question is not quite so simple as might appear. Saying that it is the corporation is not enough. The corporation is made up of far too many different people, groups, and units. Some of them have different and even conflicting interests and concerns.

The problem of corporate client identification is illustrated by the response of 850 corporate attorneys to the announcement, in late 1978, of a new government antitrust leniency plan. John Shenefield, then chief of the Department of Justice's Antitrust Division, announced to them that consideration would be given to treating the corporation leniently if the corporation voluntarily disclosed an antitrust violation by its own personnel before the government knew about it.

"Antitrust Lenience Plan Gets a Good Legal Laugh," was the way in which the legal profession's *National Law Journal* described the reaction of the attorneys present at the meeting at which Shenefield first disclosed the program. The group's sentiment was best summed up by the apparently incredulous statement of one of the attorneys that "they all want *us* to start confessing."[33] There was no need for this attorney to identify more specifically just who the "us" was. Quite obviously, most of those present at the Shenefield presentation equated individuals in the corporation, including those who might have committed antitrust crimes, with the corporation itself. This is wrong. The error must be corrected.

It may be difficult to identify just who the corporate client is, but it is not difficult to identify who it is not. The client is the corporation as an entity, not individuals employed by the corporation.[34] Perhaps, on occasion, a corporate attorney may also represent an individual employed by the corporation. However, it is becoming increasingly clear that this is unwise and improper where there is any serious possibility of conflict between the interests of the organization and those of the individual. Where there is a likelihood that the individuals may be the targets of charges of criminal misconduct, as in the cases Shenefield postulated, it is almost always sound practice for the corporate attorney to suggest that those individuals obtain their own independent legal advice.[35] Many corporations will reimburse them for the expense, without strings, so that they may be free in their selection of counsel and subsequent representation.

If a corporate employee has committed a crime in violation of corporate policy, he has jeopardized the interests of the organization. At a minimum, steps must be taken on behalf of the organization to see that he does not do it again. In addition, someone—preferably not the corporate lawyer alone—should at least consider whether or not to turn him in.[36] This is both for moral reasons and because such action will help make it clear that what he did was unauthorized and inconsistent with corporate policy. This is not by virtue of any enforcement obligation to government as such, but rather out of obligation to the corporate client.[37] Even in the absence of a "voluntary disclosure" or "lenience" policy, such as suggested by Shenefield, I have never heard of an organization being held criminally responsible for the misconduct of an employee where there was clear corporate policy against such conduct, where the corporation acted reasonably to uncover the violation, and where the individual was turned in as soon as the misconduct was discovered.[38] Rare as such cases may be, they illustrate the principle that the corporate client is the corporation, not any individual.

Unfortunately, this is not the way it usually works. Corporate attorneys, inside and outside, often have quite intimate relationships with individuals up and down the corporate personnel line, from the chairman and president on down to the lowest employee obtaining corporate legal advice. Internal corporate counsel frequently call smaller units of the organization, and people employed by those units, "clients," even though technically that is not what they are. The attorney, inside or outside, may well feel the same obligation to the organization's employee who consults him, and whom he advises, as does the independent attorney with a strictly individual client. The corporate employee similarly may treat the corporate lawyer advising him as "my attorney." He will be shocked, aggrieved, and dismayed if a confidence is violated, and a disclosure damaging to him personally is made to an outsider or even to a superior in the corporation.

This intimacy and relationship between corporate attorney and corporate employee is not just a matter of history and tradition. The corporate attorney may owe his retainer or job to an individual, or individuals, in the organization. He will not only feel a sense of personal obligation and loyalty for that reason but will fear loss of the client if the relationship is jeopardized. Where the attorney is a corporate employee, his future in the organization can be prejudiced by how other personnel feel about him. Promotions and salary increases can turn in large part upon recommendations made about him by these other corporate employees whom he has treated as individual "clients."

Some commentators have suggested that the potential conflict between organizational and individual obligations may be resolved by following Justice Brandeis' suggestion that lawyers in such circumstances should regard themselves not so much as counsel for antagonists, but as "counsel to the situation." Such a role would permit the attorney to continue his relationship with all those who have come to consider him as their counsel.[39]

The Brandeis suggestion has much to commend it in general corporate affairs. There, sound arguments can be made that practical concerns dictate continuation of a close corporate attorney–corporate employee relationship. Such a relationship, however, calls for special caution and control where socioscientific disputes are involved. There we are not simply dealing with ordinary commercial problems, but with important societal issues, in which intense personal emotions must yield to societal goals. There should be no doubt that the corporate lawyer, whether employed by the corporation or an independent outside law firm, is first and foremost the attorney for the organization and not its employees. He, and all persons dealing with him, must be made to recognize that fact and to appreciate its consequences.

Merely spelling out the obligation by saying this is not enough. Attorneys are professionals, but they are also entrepreneurs. If their retainers, salaries, promotions, and rewards generally, continue to be unreasonably dependent upon the favorable attitudes and recommendations of fellow corporate employees—some of whom may consider themselves aggrieved by the attorney's conduct—they will inevitably find it difficult to give advice or take action which is inconsistent with the interests and desires of those employees. At least in socioscientific cases, and preferably in all cases, there must be structural changes within the organization which clearly identify the corporation alone as the ultimate decision maker with regard to legal affairs. The senior responsible lawyer—ordinarily the corporation's general counsel—should be appointed by, and report to, a legal committee selected by the corporation's board of directors, preferably composed entirely of directors. All attorneys acting for the corporation, inside and out, should be under the direction and control of the general counsel, so that, through him, the legal committee and the board alone will have the ultimate corporate decision power over all legal personnel and all legal affairs.[40]

This committee–general counsel structure simplifies introduction of tight management controls in the legal area by clearly identifying authority and responsibility. By making it clear in addition that the real client is the corporation itself, not any individual or part of the corporation, this structure helps resolve the difficult personal and ambition concerns in the relationship. Even when it happens that a general counsel has a complaint about a chairman or a president, as sometimes occurs, there is a place to go.

What if the general counsel considers that the legal committee and the board of directors are acting improperly? Here the answer is certainly not easy, but the proposed structure does confirm that the corporation which is the lawyer's client is no more coterminous with its board, than it is with its lesser employees. The general counsel faced with such a structure will recognize that it identifies the corporation as a separate and independent legal entity. It is owned by its shareholders. It is directed by its board. It is managed by its officers. It is operated by its employees. It has obligations not only to each of these groups, but to bondholders, noteholders, other creditors, customers, suppliers, strangers who may somehow be affected by the company's products, and to many, many other individuals and segments of society as well. He will appreciate that an unlawful action which is voluntarily participated in by every other person in the corporation, from the most senior officer and director to the most junior clerk, is still unlawful and improper. It is in violation of the corporation's interests. After he has done what he can within regular corporate channels to change the decision, he will know that he has still further to go.

What else must he do? In some cases, a clear and vigorous protest may be enough. In others, a communication to an important stockholder may do the trick. Resignation by the general counsel may be in order. In an extreme case, where some critical overriding corporate public interest is involved, conceivably the general counsel may conclude that he must take independent action on behalf of the corporation. This may be by initiating litigation seeking instructions, or even by reporting the matter to the authorities. All of this is remote and speculative, of course. It is not my purpose to try to anticipate the details of how a corporate attorney should conduct himself under all circumstances. The important point is that these structural changes in the relationship between corporate attorney and corporate client enhance control of the socioscientific dispute-resolution function. They do this by making clear the nature of the responsibility, and to whom it extends. In any specific case, what the general counsel must do will then depend upon his legal evaluation of all the facts and circumstances in the light of that responsibility.

There are some who contend that the changes I have suggested regarding authority and responsibility are not enough. Personal intimacies, and practical realities of reward and punishment—they will say—require that the role of the corporate attorney must be just as independent of his client as the role of the certified public accountant. The CPA for the major public company is selected by the board of directors. Parts of his audits are supplied to government and to the public, as well as to management. There must be a separate corporate audit committee, controlled by outside directors; any change in CPA must be explained fully. Why should not the same be true for the corporate lawyer?[41]

Equating the role of the corporate attorney with that of the corporate CPA is unsound and dangerous. The corporate lawyer in our society is a representative or agent. He is selected solely for the purpose of advising the organization, not the public. His legal advice is "privileged." That privilege belongs to the client, not to the public. Unlike the CPA, who has clear legal obligations to third parties, the lawyer's obligation to the public is most limited, such as not to engage in fraud or misrepresentation. His assignment is to serve the corporation.

The procedures I have suggested in this chapter already cast some adversarial shadows over the corporate attorney–client relationship. The attorney may not be the corporate official who "turns in" a corporate-employee malefactor, but everyone will know that he is an "informer" for senior management. Speculation that yet untried and untested procedures will not work is no reason to add another adversarial element into a relationship which should be one of the utmost possible trust and confidence.

To return to the Shenefield proposal which introduced this section: What should the 850 corporate attorneys listening to the Shenefield announcement of the new antitrust leniency policy have done? They should not have laughed in incredulity and derision as they did. By using the word *us*, they indicated that they regarded corporate employees as their clients. They confused their responsibility to their corporations, with their relationships to individuals employed by those corporations. They should have carried the Shenefield proposal and its possible benefits back to their corporations. If the corporate policy defining the response was not already clear enough, they should have asked their corporate clients, "Should we develop (or improve) an internal antitrust monitoring program, designed to find out if anyone in the company is doing anything unlawful? If we find something, what should we do about it?"

Preventive Law

This brings us to a final aspect of change in legal management practice regarding socioscientific disputes: the corporate attorney's obligation to practice intensive "preventive law." This means education, training, investigating, monitoring, and auditing.[42]

We live in an increasingly regulated society. Environmental, consumer, civil rights, products liability, labor, antitrust, securities, and a variety of other laws and regulations are on the books, and are extremely complex and fast-moving. They are difficult for even the specialized attorney to understand fully, and keep up with. Responsibilities are great, and potential liability and exposure tremendous. As a consequence, it is already common practice in many large corporations to train nonlegal personnel in the requirements in these areas. Much, much more of this is needed, but there is nothing particularly novel or startling about the recommendation. Most corporate lawyers would readily acknowledge the need for more such education and training "if only we had the time and people!"

What is not generally recognized (except by a very few forward-looking corporations, or in a limited sense in a very few special areas, such as securities and antitrust law), is the need for corporate counsel to conduct monitoring, investigating, and auditing programs as well.[43] These would be designed to uncover pertinent corporate conduct, which—for one reason or another—might not otherwise come to the attention of the legal staff. They would be similar to the enforcement functions performed by the General Accounting Office or the Inspector General in government—and by the company's own accountants and auditors in the financial area.

More and more frequently, the press reports cases in which giant corporations, and their executives, suffer huge punitive damage awards for willful or malicious misconduct, or are found guilty of crime.[44] Yet at the same time, as referred to earlier in this chapter, senior executives are heard to protest—with considerable conviction and credibility—"I didn't know." Why didn't they know? How come a government investigator or a litigating adversary can find things out that the board of directors wasn't told about? Logically, shouldn't such information be far more accessible and available to executives in charge of the corporation than to outsiders?

Perhaps in some cases, internal personnel, including lawyers, create a more effective "cover" against colleagues than against outsiders and adversaries. But this is the exception. Ordinarily the answer to why management does not know about these things is that it never asked, or that no one ever looked. The organization never bothered to assign an enforcement function to its counsel, or failed to provide adequate resources to carry out the function in any but the most obvious areas. In addition, as indicated above, sometimes the attorneys will have confused their responsibilities to the organization with those to individuals employed by the organization. It will not have occurred to them to be suspicious of, or to investigate, their "clients." Once in a while, a corporate lawyer's primary effort will indeed seem to be to protect a colleague against discovery, even at the risk of sacrificing the corporation itself.

It may seem obvious to lay persons that the corporate legal responsibility should include the assignment to prevent or discover unauthorized and improper conduct by corporate personnel, at least in the important aspects of socioscientific dispute resolution. But it certainly is not obvious to corporate management and corporate attorneys. A good illustration of this may be found in the current debates concerning the obligation of a corporation regarding "whistle blowers."

A "whistle blower" is a person who regards his obligation to society, or to the law, or to some other nonemployer objective, as superior to that to his seniors. Daniel Ellsberg's disclosure of the Pentagon Papers is certainly the best known example, but there are many others. Whistle-blowing is a growing phenomenon, undoubtedly reflecting environmentalism, consumerism, populism, and other current social attitudes. In an earlier day, the usual response to the whistle blower was rejection, accompanied by charges of disloyalty. Today, however, there is a growing consensus that whistle-blower mechanisms must be introduced into the structure of any large organization, government as well as corporate. These would permit the whistle blower to present his case to persons outside the routine channels of authority which have blocked him thus far, on up through the hierarchy, until some satisfactory resolution is found.[45] Nowhere, however, have I seen any suggestion that the large corporation involved in major societal issues has the more fundamental responsibility of discovering the whistle blower's complaint before he has to raise it. As elsewhere in so much of dispute-resolution practice, the focus is on *response* to a problem, rather than *prevention*.

The modern large corporation is made up of thousands of employees, agents, and other representatives. It is almost inevitable that on some occasions some people will do some things on its behalf which are wrong. It is therefore just as important to have an investigative and enforcement program in the environmental and consumerism areas, as it is in the financial, antitrust, and securities areas, where the need is often recognized.

Legal monitoring, auditing, and investigating are not easy. Much of what is said and done on behalf of a corporation engaged in a socioscientific dispute will be by non-employees who are agents, representatives, consultants, or suppliers. Observing them will often require more rigorous effort than for corporate employees, because they will not have been subject to the on-going education and training programs of the corporation. The corporation's legal department should have just as comprehensive an ethics compliance program as that recommended in chapter 3 for the large corporate law firm. Outside attorneys must be supervised, to avoid the "sporting" and even unlawful conduct which sometimes results from the inevitable conflict between an instruction to "win" and an instruction to act properly.

An effective preventive law program must be designed to meet the needs of the specific organization involved. Programs will vary, depending upon a great number of factors, such as size of the organization, economics, the regulatory environment, and, of course, the risks involved. Preparing one is a multidisciplinary effort. After it is completed, installed, and working, the corporation's general counsel should be required to attest regularly to the reliability and effectiveness of the corporation's legal controls.[46] Attorneys may not welcome this additional responsibility, but it will further help to make it clear to all concerned that the lawyer's legal obligation is to the corporation alone, and not to any individuals employed by it. Instead of turning into just another burden, this task should give the lawyer much-needed independence from personal pressures which stem from the individual practice of an earlier day.

This chapter has offered a number of suggestions designed to achieve tight management and control of dispute-resolution authority in socioscientific matters. Implementation of these suggestions by way of a specific management program remains. That is not an easy assignment. As with the preventive law program, each overall legal management program must be carefully tailored to suit the needs of the organization involved. Unfortunately, as pointed out earlier, management of the legal function is rarely taught in business schools, or law schools for that matter, and there are few texts to use for guidance. To some extent, a bit of guesswork, and trial and error, is inevitable.

Some executives and lawyers believe that tight control of the legal function simply cannot be achieved. They argue that there is too much "judgment call" in law, and that outside interference will impair quality. Moreover, the contrary tradition is so strong that good lawyers will consider control "unprofessional." I most emphatically do not agree. American management techniques are generally regarded as the best in the world. They have solved far more difficult problems.[47] In addition, lawyers are representatives, who will obey the instructions of their principals if clearly given. By way of illustration, Appendix B sets forth a detailed and specific management program, designed necessarily to suit the needs of one particular company but based entirely on practices drawn from other organizations, all of which work. I urge those who have not tried tight legal management to do so before they conclude that it cannot be achieved.

As with the preventive law program, once the overall legal management program is designed, in place, and working, general procedures can be adopted to insure that the program continues to function effectively. Most significant in this connection, the client is already being held responsible by society for adversarial extremes practiced by a lawyer on its behalf. It can and should make clear to its attorney that he, in turn, will be answerable to the client, and even held personally responsible, for any "game," "trick," "obstructive tactic," or other violation of the rule of reason for which he fails to obtain specific advance clearance.[48]

RESPONSIBILITY IN SCIENCE

Science has always had trouble convincing the general public of the merits of its findings and conclusions, especially where there were differences among scientists themselves. The world was not quick to accept that the earth was round, or that humanity's place was not in the center of the universe. Such earlier debates, however, involved essentially scientific questions. They could be analyzed by application of the "scientific method." Frequently the traditional scientific "consensus" could be achieved. Ultimately, albeit sometimes perhaps even centuries later, the validity of a conclusion could be demonstrated in objective terms sufficient to convince all but the most obstinate and determined opponent—such as by a photograph of our almost-round planet televised from outer space.

In socioscientific disputes, however, science's late twentieth-century problem is not the same. It cannot be solved in the same way. One difficulty is the same lack of public confidence in the integrity of science and scientists that we find elsewhere in the socioscientific dispute-resolution process. In some respects, the concerns are greater. Scientific controversies are at the heart of certain issues which are even broader than those discussed earlier, in which lawyers and corporations are involved. These include matters such as whether the United States can effectively monitor Russian compliance with the SALT treaties; and whether the "trade-offs" between inflation, unemployment, and recession are inevitable. Society's resolution of these problems does not depend simply on more science by way of experiment and exploration, as in an earlier day, but in bringing credibility to science. We need "responsibility" in science, just as we need responsibility in the practice of law, and responsibility in corporate society. Science also must have a "public-interest" ethic.

Scientists can and should become involved with "values," and with "quality-of-life" and "public-policy" issues concerning science. Many have done so. Unfortunately, however, consciously or subconsciously, too many have also become advocates in our adversarial process. They employ the same adversarial extremes which have so damaged the credibility of other participants. Some scientists know this. They detest the excesses of the present dispute-resolution forums, with their emphases on winning alone and their *ad hominem* attacks on individuals. What do they do? They understandably respond by climbing into their ivory scientific and academic towers, and refusing to become further involved in the fray.

This "ivory tower" response is an abdication of societal responsibility. It is unfair to a public which desperately needs scientific involvement in public-policy disputes about scientific matters. It is an ineffective way to repair scientific credibility. The key reason for current concern with scientific credibility is not that scientists participate in socioscientific dispute resolution. It is that they so often confuse "quality-of-life" opinion with strictly scientific opinion. This becomes grist for the skillful adversary's mill. His attack on the societal value conclusion can easily establish that reasonable people may differ about it. This appears to be an attack on the validity of the scientific opinion as well. The scientist responsible for this confusion is seriously impeached as a result.

Sometimes a scientist's confusion of policy and science is inadvertent. Sometimes, however, it represents a deliberate and improper adversarial extreme. The latter is the case when an effort is made by "adversary science" to cloak value judgments about which people may quite reasonably differ with the mantle of a scientific expertise which the lay person is not qualified to question. In either case, it is wrong because it is misleading, and because it seeks to set scientist and science above the rest of society.

Science and scientists should not withdraw from the public struggle. Instead, they should separate and clearly distinguish between "science" and "value," or "quality of life," or "public policy." They must then communicate their science to the public in as effective a manner as possible, so as to make it very clear that they want (and intend) the public to participate as full partner in the decision process. This chapter deals with these matters.

Science and Public Policy

The fine line separating "pure" science from "value" or "public policy" is most difficult to define conceptually. In addition, it is often very hard or even impossible to find practically. A simple, all-too-common hypothetical example of an easy case will help explain why this is so.

In recent months, you have been experiencing pain after eating a heavy meal, when you engage in some physical activity, such as walking up a flight of stairs, or when you are excited. Usually the pain is located over your heart, in your chest; sometimes it is in your shoulder or neck; sometimes in your left arm. The pain is one of constriction and suffocation, but is not sharp. Often it is accompanied by dizziness and palpitations. Attacks are usually short, with the worst of it over in a few minutes.

At your regular annual medical checkup, your doctor tells you that you have all the classic symptoms of *angina pectoris*. He prescribes nitroglycerin for relief but, of course, it will not correct the underlying problem. He tells you to avoid excitement and unnecessary exertion, to go on a low fat diet, to give up smoking, and to avoid taking too much coffee. He doubts that there is a great deal else to be done to get you back to where you once were, but recommends that you consult a cardiologist. He says that perhaps a specialist in these matters can pinpoint the cause of the problem and offer other alternatives.

Dr. Jones is a distinguished cardiologist and heart surgeon. He has a very busy practice, and you must wait a few weeks for an appointment. When you arrive, his office is crowded with other heart patients. Finally your turn comes, and you are given a complete physical examination, including an electrocardiogram and a full series of X rays. When all this is completed, you find yourself in Dr. Jones' office, sitting across from his desk, anxiously and nervously awaiting his diagnosis, prognosis, and advice as to what you should do.

Dr. Jones comes in from an examination room to your right, having just seen a patient. There is another patient in the examination room to your left, pacing up and down, waiting to be seen. He seems just as nervous as you. Dr. Jones sits down, examines the charts and papers on his desk very carefully and, after what seems much too long, looks up. He says, "I think you'll be very glad to know that we can correct your problem. There's every indication that a heart bypass will do the trick." He says a good deal more, but you are shocked, feel pressed and rushed, and do not really seem to hear all of it. You do remember, however, asking him for his recommendation at the end of a few minutes conversation, and his telling you, "There's no doubt in my mind that you should have open heart surgery." He promises to send his report to your internist immediately.

Two days later you telephone your internist, who has received Dr. Jones' report. He says that it is a good report, and that he has no real questions about it but, as a matter of caution, he always believes in getting a second opinion when there is any real question as to how to proceed. The recommended operation is a serious one and perhaps other approaches are possible. You make a second appointment, with Dr. Smith, also a distinguished and very busy cardiologist, and go through an almost identical routine. This time, however, he tells you that he does *not* think open heart surgery is indicated. He has patients who have lived with your condition for thirty years and more, and who are in better health than he is. He recommends that you follow the original advice your internist gave you about diet, smoking, stress, and exertion. If you do so, his experience is that you'll outlive many of your friends who have no apparent heart problem at all.

What has happened here? Are these two cardiologists seeing different things and drawing different conclusions about your condition and its treatment? That is possible, of course, but yours is a very simple case, and the much greater likelihood is that they are really expressing value judgments about the way *you* should live. If you do not undergo surgery, you must avoid exertion and excitement and watch your diet carefully. This means that you cannot play tennis, climb mountains, eat gourmet meals, or speculate heavily in the stock market. If you do have surgery, there's an "excellent" (85 percent?) chance that you will be able to live as you once did, or at least much better than otherwise. The operation will cost a great deal of money (do you have surgical insurance?), keep you out of commission for a good deal of time (will personal or business life permit?), and involve some risk. There is "some" (15 percent?) chance that your condition will not be improved; a "small" (5 percent?) chance that your condition will get "worse" (how bad?), and a "slight" (1 percent?) possibility that you will not survive the operation. Both Drs. Jones and Smith, and your own internist, have told you all this if you examine their words very carefully. But we are all victims of our own perceptions, and the ways in which they phrased their words, and the tones and emphases, reflected their own judgments of "quality of life." Without going into all the details of why each feels as he does, it is Dr. Jones' honest and considered belief that you should not be forced to live so restricted a life when the risks are so slight. It is Dr. Smith's equally honest and considered belief that the restrictions on your life are not really very serious ones at all and certainly are not worth the risks involved in the operation.

I have deliberately postulated a case in which the "quality-of-life" judgments inherent in a serious decision as to surgery did not get through to the patient. Often, of course, surgeons are able to insure that patients fully understand and appreciate everything that is involved. But good doctors are very busy these days. In addition, telling a patient of the risks can be a highly personalized affair. Surgeons sometimes seem to believe that they should avoid such personal exchanges, lest any resulting emotion somehow interfere with the cold judgment essential to the wielding of a surgical knife. Many lawyers consider that the objectivity and impartiality of legal judgment suffer when they represent persons with whom they have close personal relationships. Many surgeons refuse to operate on close friends and relatives for similar reasons. Refusing to accept a patient is one thing, however. Once the patient obligation is accepted, excessively dispassionate treatment of the patient is quite another. It can stand in the way of effective patient communication of emotional concerns. It can exacerbate the patient's decision-making problem.

The inevitable consequence of a failure to separate out, and clearly distinguish, true medical opinion from "quality-of-life" judgment is to communicate a conclusion which is surreptiously (albeit innocently) laden with value decisions. This precludes effective decision-making by the patient. I believe this failure adequately to communicate is an important cause of medicine's current serious malpractice problem. There are other causes, of course, including true medical incompetence and attorney barratry. But even without such incompetence or barratry, a patient may blame his doctor for malpractice because the patient did not really understand the risk to which he was being subjected. A skilled doctor telling a patient that there is a "slight" (1 percent) risk of harm from surgery, knows that in every one hundred cases, on the average, one patient will suffer. This patient may be the one. The patient, however, may translate "slight" into "insignificant," or "trivial." However convinced the doctor may be that he has communicated the risk, this patient knows that the patient did not factor it into the patient's decision-making, and did not intentionally assume the danger involved.[1]

Separating Science and Policy

The scientist's problem in the socioscientific controversy is different from the surgeon's only in degree—it is far more difficult to correct. When a toxicologist says, "The risk is acceptable," he is stating a conclusion that includes toxicological data, research, and opinion, about which he *is* specially qualified, and "value" or "public-policy" conviction, about which he is no more qualified than lay persons.[2] As I have said, he has a right and a societal duty to express both views. But when he confuses the two so as to make it appear that he has some special right to decide "quality-of-life" matters for others, he evokes precisely the same response as the medical practitioner who has usurped that function. He exposes himself to the most devastating cross-examination. This commonly compels the admission that he has erred, and the inference that he has lied. He becomes one of the adversarial extremists in the "battle of experts" which so often impeaches the whole of science.

The medical/surgical hypothetical example given above, was intentionally, although perhaps simplistically, selected to illustrate the difference between "pure science" and "value." In practice in the real world, however, the distinctions between the two are often blurred and extraordinarily difficult to decipher.[3] Sometimes classification as "science" or "value" will depend on the ultimate issue under consideration. Sometimes what at initial blush seems to be value alone will turn out to have an important scientific ingredient as well. Sometimes exploration in depth of something which at first appears to be science may reveal an important value component.

For example, a research scientist involved in a controversial dispute may subconsciously wish to be on the "safe" side. Depending on the nature of the anticipated adversarial attack, he will include or exclude "suspicious" tissues within his "carcinogenic tumor" count. His conclusion as to whether the evidence of carcinogenicity is or is not "statistically significant," then, will have been made to turn upon a "judgment call" on an issue in scientific question, rather than upon the kind of positive evidence which is implied by the terms "scientific facts" and "science."

Moreover, unlike the medical/surgical hypothetical example above, the scientific issues in socioscientific disputes are usually at the leading edge of science, where there is ample room for serious scientific doubt. It may surprise the lay person, but pathologists will sometimes debate long and hard as to whether tumor cells examined under an electron microscope are benign or malignant. Similarly, analytical chemists will argue heatedly whether a low-level gas chromatographic signal represents one or another chemical, or is simply background "noise." In speaking with each other, each will understand what is meant by phrases such as "there is substantial risk of malignancy," or "there is no evidence of serious contamination." It will not occur to them to explain that "substantial" means "more than a 10 percent chance" (assuming that to be the recognized test here), and that "no evidence" means "below a signal-to-noise ratio of two or one." They all know it.*

The first step in effectively separating scientific data and opinion from value and policy, then, is recognition of the need to do so.

* Even trial lawyers and trial law professors who have not been involved in socioscientific disputes assume that these kinds of "scientific facts" are not ordinarily in dispute in such matters. This misimpression has sometimes led to unsound recommendations for revision of the dispute-resolution process.

Scientists must have the distinction prominently in mind as they communicate to lay persons. They may not pretend expertise in nonscientific matters. They may become adversaries, but they must *not* practice adversarial extremes. They must understand that they have just as much responsibility to society in the resolution of socioscientific disputes as any other party, including the lawyer.[4] They must resist the efforts of advocates in the adversarial process to persuade them to do what they know they should not do. No matter how much an employer or an advocate wants a firm, positive, dogmatic conclusion, their involvement must be that of the "balanced, objective scientist who scrupulously qualifies his statements to guard against overinterpretation."[5] If that renders the scientist useless, or gets him fired, so be it.

Professional scientific societies have an important role to play in this area. Codes of ethics should not only mandate adherence to "scientific method," but prohibit the employment of adversarial extremes in socioscientific disputes. The rule-of-reason approach can be spelled out as a guideline to dispute-resolution conduct. Such provisions would not only help in the educational and "consciousness raising" process, but would furnish scientists a quick response to those who urge adoption of contrary practices.[6]

Professional societies can also play an aggressive and affirmative enforcement role where adversarial extremes threaten to denigrate credibility, and to bring the profession into disrepute. They need not stand idly by and watch such tactics. I was present during a nuclear-plant licensing proceeding when the hearing board received a request from a special panel of a professional society to testify following the appearance of one expert. The special panel was made up of three distinguished scientists, who appeared strictly on behalf of the society, and at their own expense. They read a resolution adopted by the society which condemned the misleading, "nonscientific" conclusions of the expert who had preceded them, and who had been testifying to the same effect in a host of other cases. Thereafter they went on to identify the "value" judgments which the expert had made, chapter and verse. Not only did the panel's testimony contribute greatly to the hearing board's understanding of the scientific issues, but it demonstrated just how "professional" scientists can be. It did much to enhance the credibility of science in that case.

I am not suggesting that professional societies, or any other institution of science, should interfere with responsible debate about scientific matters. Societies have no right to censor or suppress what scientists say or do in public-policy disputes—where individuals act as scientists. Those are matters on which society at large must exercise the final critical role. But professional societies, and science, can and should lay out guidelines for what is ethical, what is professional, and what is "scientific." It is up to them to stop the practice of "adversary science" and thereby bring the "scientific method" into socioscientific dispute resolution without reservation.[7]

Science and Public Communication

Credibility in science requires also that science and scientists do what they can to help the lay public understand scientific matters. There is always some possibility that even the most carefully considered "scientific" statement will also contain a measure of "value." That possibility is minimized where a real effort at communication is made. Whatever confusion may still occur is seen to be clearly innocent in any event and not the result of an "adversarial" effort to mislead. In addition, at least for the present, scientific credibility is not secure. Serious efforts at explanation instill confidence.

A vital first task, on the "macro-science" level, is to insure that there is general recognition of this need for communication between science and the public. Some steps in this direction are under way. In October, 1975, U.S. Senators Edward M. Kennedy and Jacob K. Javits wrote a letter to Willard Gaylin of the Institute of Society, Ethics and the Life Sciences, and to Frederick C. Robbins, dean of Case Western Reserve Medical School. They requested that a conference be held to consider the relationship between science and the public. *Science* magazine reported their view as follows:

> The senators and members of their staffs had been disturbed by their perceptions of great tensions between the scientific community and Congress and the public, as well as within the community itself. "As legislators who have long been concerned with national health problems, and particularly the support of basic biomedical research, we find disquieting both the tone and direction of some of the current discussion about the public role in the establishment of science policy," Kennedy

and Javits wrote in their letter proposing the conference. "There are suggestions that the scientists may not appreciate adequately the public interest and its role in decision-making," they said with masterful understatement, while acknowledging on the other side that "the public may not adequately understand the scientist."[8]

The result was a three-day meeting, at which some sixty Nobel Prize winners, lawyers, legislators, anthropologists, journalists, and others considered the problem presented. Not surprisingly with such a large group, there was no clear decision at all as to just what specific steps should be taken to remedy the problem. There was agreement, however, that the problem is a serious one, and requires attention; the sponsorship and participation suggests that these conclusions will be pursued.[9]

Science reports that "it seems certain that the movement for public participation in science is gaining strength."[10] The New York Academy of Sciences, one of the first scientific societies established in the United States, has begun a program to "increase its involvement in public policy issues and to help evolve a better—and it feels, long overdue—public understanding of science and scientists."[11] There is growing awareness of the need to improve, expand, and hasten these initial efforts. Shortly after the Three Mile Island nuclear episode, in a news analysis captioned "Scientists and Society's Fears—Nuclear Accident Aggravates the Public's Distrust of Assurances From Experts on Technical Topics," the *New York Times* reported:

> The nuclear accident that occurred at Middletown, Pa., has raised questions not only about the future of atomic power but also about the reputation of the scientific and engineering communities. How much trust, it is being asked, should a nonscientific public place in the pronouncements of the scientific community? What are the limits of technical knowledge?
>
> "The disbelief in experts was already strong enough," says David Riesman, Harvard sociologist, "but it now has been given an enormous boost. . . ."
>
> At its heart, the problem is one of how scientists can constructively communicate lack of knowledge in a high-risk situation. "How do you know what you know, and what are the limits to your knowledge?" asks Alan McGowan, president of the Scientists' Institute for Public Information. "When you are putting the public safety on the line, knowing that

> there are bound to be gaps in what you know, how do you communicate this to society? . . ."
>
> "What worries me most and is a source of some pessimism is that science and technology ought to be an integral part of the culture, and if there is a sense of forward momentum the society has to share in the risk," says Paul Doty, Harvard biochemist and director of the Center for Science and International Affairs at the Kennedy School of Government.[12]

Leon Golberg, a past-president of the Society of Toxicology (SOT), first President of the Chemical Industry Institute of Toxicology (CIIT), and eminent toxicologist, has proposed that SOT conduct an in-depth analysis of this same problem as it arises in toxicology.[13] This also is only a beginning. However, it adds to the increasing body of evidence that more and more scientists and their institutions, and society generally, are coming to understand the obligation of science to communicate effectively with the public. With determination and commitment, they can move forward with the assignment, full steam ahead. Socioscientific decision-making will be much the better for it.[14]

Simplifying Complex Information

"Macro-science" recognition of the need to communicate between science and public is important. So is actual, "micro-science" communication itself.

Almost by definition, a lay public can never fully understand and evaluate all of what science has to say. Were it otherwise, there would be little to justify the Ph.D. and years of experience required of scientists involved in socioscientific disputes. Given the necessary assistance, however, the public can understand much of science, just as the lay patient can understand much of what the surgeon says and does.

The lay mind works in a peculiar way with regard to complex scientific matters. If it has understood and grasped a part of the technical language and some of the esoteric considerations, it extrapolates to the balance and accepts the whole.[15] A patient may have a most informative explanation from his surgeon, even if it is also necessarily brief and limited. Just listen to the patient then talk about "my operation." He may accurately repeat some of the simpler specific things the surgeon said, but he will also go on with all kinds of inferences and conclusions. Some of the latter will be totally erroneous. In this case that makes no great difference. The patient had no real need for the additional information in connection with his decision-making process. The important thing is that the patient is satisfied and happy.

Where possible, the scientist must always translate scientific terminology into lay language. Whether deliberately or not, the doctor who uses Latin terms to describe an organ, a condition, a disease, or a drug, impairs the patient's ability to understand. Sometimes there is no fully equivalent lay term, and the closest word may be misleading. But usually there will be something one can do. The effort to explain all by itself, honestly undertaken, will enhance credibility.

More than just words must be translated. Scientific concepts must be communicated in terms which the lay person can understand. The concept of "trace concentration" provides a good example. These are the small amounts of substances found in air, water, and elsewhere in the environment, which may create an environmental hazard. Arsenic in drinking water and asbestos fibers in the air are illustrative. Scientists speak of them in terms of "parts per million" (ppm), "parts per billion" (ppb), or "parts per trillion" (ppt), to indicate that there is one part of the contaminant (arsenic, for example), to every million (ppm), billion (ppb), or trillion (ppt) parts of the carrier (water).

Modern analytical chemistry has been greatly improving its ability to detect such trace concentrations in recent years, to the point where under some circumstances and with some chemicals, it can detect concentrations at even less than the ppt level. As other scientific specialties learn more and more about the effects of chemicals at these levels, there is increasing demand for lower-level detection. Analytical chemistry, accordingly, must keep ahead even of the other sciences which are at the scientific forefront in environmental disputes.

In such a rapidly developing discipline, it is little wonder that there are often very strong differences of opinion among analytical chemists. One, using so-called "thin-layer chromatography," for example, will conclude that he "sees" a signal which is that of the particular compound for which he has been searching. A second, examining precisely the same chromatogram, will disagree. His opinion will be that the suspect signal is not sufficiently different from a whole host of other signals which are "noise"—the routine, background signs of chemicals which are always present in the environment. Where the presence of the compound must be evaluated in a socioscientific dispute, these two scientists will both testify as to what they have seen, and to their conclusions. Because there is no scientific agreement or consensus, the issue must be decided by the arbiter of the dispute and, if the controversy continues, finally by the lay public.[16]

Of course, the public will not be able to understand all of what each of these chemists says and does. It can be made to appreciate, however, just how difficult it is to find concentrations at these extremely low levels. This, in turn, may make it possible to reject the "absolutist" approach of one of the chemists as incredible, because he asserts his findings in unqualified terms. It will thereby make it possible to accept the assertion of the second chemist, who has expressed his opinion in terms of probabilities ("A 60 percent chance that it is chemical A, a 20 percent chance that it is chemical B, and a 20 percent chance that it is simply noise.")

How can the public be made to understand the difficulty of finding these trace concentrations? One way is to translate trace concentrations of chemicals, which are the issue, into trace concentrations of other things which are common to the public experience, and which the public will therefore understand. Warren B. Crummett, a distinguished analytical chemist who has pioneered in his specialty, has pioneered as well in this kind of communication. He has developed a chart of popular trace-concentration equivalents, which I have seen him use effectively and persuasively with lay arbiters and lay audiences (Table 6.1). It is designed to reach some interest and therefore the consciousness of almost anyone, whatever the viewer's or listener's special preference or prediliction. It thereby also convinces the audience that a real attempt at lay explanation has been made.

Nuclear physicist and Nobel Prize-winner Hans Bethe has used a similar type of visual aid to compare nuclear power plant hazard

Table 6.1 Trace-Concentration Units

Unit	1 ppm	1 ppb	1 ppt
Length	1 inch/16 miles	1 inch/16,000 miles	1 inch/16,000,000 miles (A six-inch leap on a journey to the sun)
Time	1 minute/2 years	1 second/32 years	1 second/320 centuries or 0.06 seconds since the birth of Jesus Christ
Money	1¢/$10,000	1¢/$10,000,000	1¢/$10,000,000,000
Weight	1 ounce/31 tons	1 pinch salt/10 tons potato chips	1 pinch salt/10,000 tons potato chips
Volume	1 drop vermouth/ 80 "fifths" gin	1 drop vermouth/ 500 barrels gin	1 drop of vermouth in a pool of gin 43 feet deep covering the area of a football field
Area	1 sq. ft./23 acres	1 sq. in./160-acre farm	1 sq. ft. in the state of Indiana or 1 large grain of sand on the surface of Daytona Beach
Action	1 bogey/120 golf tournaments	1 bogey/120,000 golf tournaments	1 bogey/120,000,000 golf tournaments
	1 lob/1,200 tennis matches	1 lob/1,200,000 tennis matches	1 lob/1,200,000,000 tennis matches
Quality	1 bad apple/2,000 barrels	1 bad apple/2,000,000 barrels	1 bad apple/2,000,000,000 barrels
Rate	1 dented fender/10 car lifetimes	1 dented fender/10.000 car lifetimes	1 dented fender/10,000,000 car lifetimes

with other common hazards. He portrays a pie, simply divided into slices proportional to each hazard. His statistics may be subject to change today, because of revisions of the calculations in the so-called Rasmussen report, dealing with nuclear hazard. However, the usefulness of Bethe's communications approach is unaffected by any changes in the figures. His pie is divided as follows:

Hazard	Number of Deaths
Motor vehicle accidents	50,000
Other accidents	30,000
Falls	18,000
Fires and hot substances	7,000
Drowning	6,000
Routine emission of radiation	2
Nuclear reactor accidents	2[17]

It bears repetition that what is science and what is value may depend on the ultimate issue under consideration. Communication aids, like statistics, can be designed to support contentions and to serve as adversarial extremes. The Crummett chart is a valid effort to show that trace concentrations are extremely difficult to identify. It is *not* valid to show that trace concentrations are necessarily acceptable, or "safe." Yet some persons have sought to use the chart in that way. In similar fashion, an American Cancer Society epidemiologist is quoted as saying:

> There are a few cancer-causing substances that seem to be dangerous even at levels of one part per million. The aflatoxin sometimes found in peanut butter is such a substance. But one part per *billion* of anything seems just too small to worry about.[18]

Of course, whether or not something is harmful at the ppb level—or even at the ppt level or below—depends upon its toxicological effect, not its dimensions. 1 ppb is 1,000 ppt. For certain substances under certain circumstances, it can kill. To suggest otherwise is to mislead.

The same kind of analysis applies to the Bethe hazard "pie." It is valid to support a contention that society accepts some risks which appear to be far greater than the nuclear risk. It is not valid to support a contention that small risks are acceptable. That is because it contains no reference at all to the benefit side of the equation, which is the only basis on which society accepts any risk. Yet consider the following paragraph taken from the Bethe article. It might be interpreted as suggesting that an argument is being made that nuclear risks should be accepted because they are small:

> It is obvious that 5,000 cancer deaths [from nuclear disaster] would be a tragic toll. To put it in perspective, however, one should remember that in the U.S. there are more than 300,000 deaths every year from cancers due to other causes. A reactor accident clearly would not be the end of the world, as many opponents of nuclear power try to picture it. It is less serious than most minor wars, and these are unfortunately quite frequent. Some possible industrial accidents can be more serious, such as explosions and fires in large arrays of gasoline storage tanks or chemical explosions. The danger from dam breaks is probably even greater.[19]

Standing alone, which Bethe undoubtedly did not intend, this presentation invites sarcastic repetition of the term "minor war," and the adamant statement, "Our society will not voluntarily accept even *one* case of voluntary murder!"

The rule of reason forbids tactics of this kind which, intentionally or not, are calculated to mislead. All those who participate in socioscientific disputes, and aim at serving the public interest, must conduct themselves in accordance with the philosophy of the rule. Scientists are no exception.

Apples and Oranges

Effective communication of scientific information to the lay public also requires that scientists present comparisons which are in the same terms, so that they can be balanced and evaluated by lay people. At present, especially in presenting the "risk/benefit" analysis which is so critical to socioscientific dispute resolution, science sometimes offers us only apples to compare with oranges: we are led to believe that quality of life, health, and even survival—one side of the equation ("apples")—are to be equated with dollars alone on the other side ("oranges"). Little wonder that so many lay people shudder at the prospect.

I use the phrase *risk*/benefit deliberately, but as I have already pointed out, the more common one is *cost*/benefit. Undoubtedly that is because the typical, careful scientific analysis of an environmental issue focuses so much of its concern on "cost" expressed in strictly economic terms. One side of the equation will be stated in terms of real human concerns, such as cancer, genetic mutation, death, destruction of the environment, and other horrible consequences, which quite understandably evoke the most intense adverse reactions. The other side, however, will be phrased in money or equivalent economic terms—dollar cost to industry, dollar reduction in foreign trade, contribution to inflation, earning power lost, etc. It does not matter which side of the equation is which, pro or con, risk or benefit. The inevitable, unfortunate public conclusion is that an effort is being made to equate dollars with lives, a callous, macabre, terrifying approach. Thus, the *New Yorker* defines the risk/benefit assessment process as "the careful weighing of the *economic* against the environmental and public-health considerations involved in the proposed regulatory solution."[20] Senior Federal Circuit Judge David Bazelon, who should know better by virtue of long experience as an arbiter in this field, refers to "the trade-off between dollars and lives."[21] A report in the *National Journal* puts it this way:

> Whether or not the regulators like—or find it politic—to admit it, the economists insist that all decisions in the health and safety area implicitly put a value on life. A staff member of the White House Council of Economic Advisers (CEA), who asked not to be identified, laughingly referred in an interview to the published work of one pair of academic economists who have set a precise value on a life of $255,000.
>
> The CEA economist went on to explain that few are willing to deal so arbitrarily with the issue. "But it stands to reason," he continued, "that there is a range of expenditures that we would reasonably expend to save a life. For example, I would agree the general public should pay $10,000 to extend anyone's life, even if just for a short while, but I would not agree that they should pay $100 million. Somewhere in between, a line has to be drawn."[22]

A *New York Times* article phrases it even more brutally:

> First, cost-benefit tries to equate what is economically, let alone morally, unequatable: kilowatt-hours of electricity produced by a nuclear power plant versus the number of cancer deaths caused by a slow radiation leak at the site where the spent fuel is stored, for

example. This kind of equation makes no sense unless we can decide how many kilowatt-hours of electricity a human life is worth.[23]

Any economist knows that money is not important in itself. Try to do something with it on a desert island. Money's significance is as a medium of exchange, making possible the allocation of limited resources. With very rare exceptions, air is everywhere, freely available to all. No dollar value is assigned to it. In some places, the same is true for drinking water; in others (remember the Ancient Mariner's "Water, water, everywhere, nor any drop to drink"?), pure water is scarce, with a high dollar value. Kidney dialysis equipment, in contrast, is almost always very hard to come by. It requires time, skill, labor, capital, and raw materials to produce. A dollar value is assigned to it, as a way in which society decides who will have it and who will not. At any given point in time, there are only so many pieces of this equipment to go around. Thus, when we assign them to some people, we withhold them from others. If we decide to produce more of them, we must similarly allocate scarce metals or skilled labor or something else from another activity, such as radiation therapy. When we do so, we deny someone else the benefits of *that* treatment. Until the day when all resources are as freely available as air, the allocation of something to one activity or person will necessarily mean that it is denied to something or someone else. This may be an oversimplified discussion of economics, but it is essentially the core of the risk/benefit trade-off. Robinson E. Barker, chairman of PPG Industries, described it as follows:

At any given time, we have a finite quantity of resources available to devote to the satisfaction of our needs and desires. We may shift resources from one sector of the economy to another, but the total of claims on our resources cannot exceed the total of the resources themselves. We should not delude ourselves and our fellow citizens that some problems can be dampened without kindling other problems. In such areas as unemployment versus inflation or a cleaner environment versus higher productivity, there must always be trade-offs.[24]

Science must express both sides of the risk/benefit equation in terms of the most common types of concern possible. Often the terms can be identical—lives lost can be compared with lives saved; health impaired with health improved; natural beauty destroyed with natural beauty saved. Dr. Wayne Binns, a senior government toxicologist, brought this home to me many years ago while we were discussing the risk/benefit aspects of a common herbicide used to control weeds. I routinely was comparing the toxic effects of the herbicide with its benefits in terms of greater food production in areas of acute starvation, certainly a valid trade-off even if the terms were not identical. He added a far more telling comparison: the toxic effects of the weed seeds eliminated by this herbicide were far worse than those of the herbicide itself—*not* using the herbicide resulted in more toxicity than using it. In similar fashion, we can compare the hazards of nuclear power with the hazards of not having nuclear power, including the hazards of coal mining, the hazards of operating coal-fired plants, the hazards of relying on Middle East oil, and the hazards of inadequate heat, light, or power. Comparison of important human and societal values with dollars alone is misleading and biases the public response, just as it did the *New York Times*.

As with the use of other scientific-communication aids, however, there must be caution lest this kind of analysis itself be misused. It has been contended, for example, that "the removal of the smog of Los Angeles would, according to present theory, lead to a doubling of the present rate of skin cancer in the affected area—just because of the added sunlight it would let in."[25] Whether this statement is technically true or not, it may mislead the lay person. It uses the terror-charged word *cancer*. It elides the underlying questions of "Just how serious *is* the present Los Angeles risk of skin cancer in comparison with the risks incident to the Los Angeles smog?" and "How serious is *skin* cancer itself in the first place?" Quite probably there could be no really serious comparison were it not for the fear and emotion which the word *cancer* inevitably generates in almost any connection. Any perceived effort of this kind to mislead will surely impair rather than repair scientific credibility.

Standards and Benchmarks—the Risk Index

Another important aid to scientific communication to the public is through the use of "standards" and "benchmarks." These can make it possible for the lay person to compare esoteric matters—and new, strange, and foreign risks—with things that are common and personal, and that lay people can therefore understand and evaluate.

One of the conclusions of the NAS Dispute Resolution Forum Workshop referred to in chapter 5 was the following:

> The scientific community must intensify its efforts to enable the public to understand, or at least to evaluate and judge, scientific information so that public consensus or accommodation is possible. Reference in this connection was made to the usefulness of standards, professional accountability, certification of procedures, codes of ethics, and other specific measures in the effort to enable the lay public to deal with esoteric technical matters. The concerned lay person who traveled here all the way from Montana because of concern about the quality of life in her home community must be as satisfied as the most technically qualified energy expert. The public must be treated with dignity and respect not patronizingly or with public relations gimmickry. (Appendix A)

The public should be given the ability to compare the qualifications of individuals and of institutions, equipment used, test methods, and whatever else may be the subject of controversy when a socioscientific dispute arises. Although there have been starts in some of these areas, such as the government's publication of so-called "good laboratory practices," they have not yet come to be generally accepted as standards against which the scientific community should measure all of its efforts. Even the definition of "toxicologist," the scientist at the heart of most hazard assessment, is imprecise and uncertain. Few institutions yet give degrees in the science of toxicology. How in the world, then, is the public to be able to compare the qualifications of one such expert with those of another? We need a major effort by scientists, professional societies, and scientific institutions generally to help us understand the comparisons.

Robert L. DuPont, director of the National Institute on Drug Abuse, has complained of the inadequacy of our present understanding with regard to these kinds of comparisons, especially as to risk. In a *Science* magazine interview, he said:

> We have, as a nation, intellectual scales that we'll put issues into to try to weigh them. The scales on which we put recreational drugs are quite unique. They go something like this, as far as I can understand it: If 20 or 30 percent of the people who use a substance do not die within 1 year, it is a pretty safe substance and, therefore, we say that a person should

> have freedom of choice about whether or not he uses it. If he doesn't hurt anybody but himself, it's up to him to decide. Then we've got another scale that the people at FDA (Food and Drug Administration) just down the road have for pharmaceutical products. And that scale says that if three mice out of a thousand die of cancer when exposed to a substance, well, heck, you pull the darn thing right off the market. Never mind having any evidence of human toxicity whatsoever. We don't give people a choice of whether they want to use cyclamates in the United States. And I find this disparity between how the same people—liberal, right-thinking, progressive souls—in our country want to relax our prohibition against marihuana and, at the same time, increase our prohibition against cyclamates or red dye No. 2 or birth control pills, to be intellectually curious at best. I think we've got to get the two scales closer together before we can be realistic on either side.[26]

Baron Rothschild, a leading British zoologist and former director of the British government's Central Advisory Council for Science and Technology, has proposed the "risk index," which would address DuPont's complaint. He phrased it this way in a November 1978, BBC television address:

> There is no such thing as a risk-free society. Even a *virtuous* life has its risks, as illustrated by the Chinese proverb: "The couple who go to bed early to save candles end up with twins." But there is no point in getting into a panic about the risks of life until you have *compared* the risks which worry you with those that don't—but perhaps should. Comparisons, far from being odious, are the best antidote to panic. What we need, therefore, is a list or index of risks and some guidance as to when to flap and when not.[27]

Richard Wilson, a physicist and acting director of the Energy and Environmental Policy Center at Harvard, has furnished one example of what such a risk index might look like. He has "calculated" the "level of exposure that increases the chance of death by 1 part in a million (or, expressing it another way and more precisely, that reduces one's life expectancy by 8 minutes)." His figures are:

Smoking 1.4 cigarettes.

Living two months with a cigarette smoker.

One X ray (in a good hospital).

Eating 100 charcoal-broiled steaks.

Eating 40 tablespoons of peanut butter.

Drinking 1,000 24-ounce soft drinks from recently banned plastic bottles.

Drinking 30 12-ounce cans of diet soda containing saccharin.

Living 20 years near a polyvinylchloride plant.

Living 150 years within 30 miles of a nuclear-power plant.[28]

As Wilson's table suggests, the method of computation or choice of examples can "load" any comparison, and bias its use. It is therefore of great importance that work of this kind be undertaken in a kind of "controversy vacuum." It should be completed well before the dispute in which it is to be applied is even in contemplation. Properly administered, this kind of "risk indexing" can be made fair and impartial, and therefore most useful. The lay public readily accepts certain risks which it understands, and where it considers the associated benefits to be worth the hazard. Obvious examples are automobile driving, where (depending upon the maximum speed fixed) we have an almost guaranteed 40,000–50,000 deaths per year in this country; and cigarette smoking, now widely accepted to be a potent cause of lung cancer. There are many, many other examples, including some which people take quite deliberately *because* of the hazard involved, such as "sky diving."

For most of the hazards incident to socioscientific disputes, however, the risk is unknown, not voluntarily assumed or controlled in the same sense as cigarette smoking and driving are, and often is at first expressed in words which are incomprehensible to the lay person. Rodent tests may indicate that a given compound, such as saccharin, has a potential for causing certain kinds of cancer in certain species of rodent. Rodents are different from people, however. Conclusions as to the resulting *human* risk, therefore, must be expressed in terms of statistical extrapolations. All of this is foreign to the public. When the dispute gets heated, the adversaries may begin to use comparisons, such as "no worse than smoking," or "more dangerous than mustard gas," but these are understood to be presented in the standard context of our adversarial system, and usually receive little credence as a result. In the summer of 1978, for example, Morris N. Cranmer, director of the Food and Drug Administration's National Center for Toxicological Research, completed what he termed a "final report on saccharin," designed to put the saccharin cancer risk in perspective. An industry trade journal summarized some of his findings as follows:

> Cranmer writes in the report that the cancer risk from consuming a saccharin-containing diet soft drink may be less than the cancer risk from an equivalent amount of sucrose (table sugar).
>
> Specifically, Cranmer asserts that the equivalent "sweet dose" of sucrose carries 375 times the "risk of cancer in animals" of a 12-oz. diet soda. He adds that a "typical steak has approximately 20 times the bladder cancer potentiating effect (if estimated from rat data) of a can of diet soda." And he notes also that "five salted peanuts would about equal the risk of a can of diet drink."[29]

It is not in derogation or deprecation of Cranmer's important work to report that it had little if any public impact, whatever may have been its significance in scientific circles. In the trial lawyer's language, it was published *post litam motam*—after the dispute arose, when perceptions are suspect. Had there been a preexisting risk index available when the saccharin problem first arose, into which the saccharin risk could be slotted with the same kind of comparison, the public reception might have been different.

I do not suggest that properly administering the risk index to avoid subtle unfairness in selection of comparisons or other bias will be easy. But surely it is well within existing scientific competence to develop common ways of comparing and "rank ordering" a variety of the important different risks of present concern, perhaps by way of sliding scale or chart. Thus, various rodent tests might be conducted to determine the risks of rodent cancer resulting from exposure to stated amounts of radiation, cigarette smoking, aflatoxin (found in various grains, peanuts, and peanut butter), vinyl chloride (used in the manufacture of phonograph records, upholstery, and a great number of other consumer products), and other common substances. These risks could then quickly be compared with whatever new develops, and well before any dispute becomes heated. The value of this kind of information to the public would be enormous.

These suggestions with regard to scientific standards and benchmarks obviously are but a beginning. Undoubtedly there are many other ways in which scientific effort can enhance public understanding.[30] Moreover, the nature of the controverted issues will inevitably change over the years. As it does, new and different scientific approaches will be required. What is needed most at this point is recognition that science and scientists must communicate with the public credibly and effectively, in scientific, nonadversarial fashion.

The "Establishment" Role

The search for more and better knowledge is a never-ending quest. Differences of view among people with different backgrounds, statuses in society, moralities, religions, resources, education, and values generally, will be perhaps inevitable until everything is known—a day that can never come. In the meantime, it seems inevitable that in their own self-interest, those who have the abilities and resources—the "establishment" in modern slang—must hold the laboring oar. This was the conclusion of the NAS Workshop which considered the matter (see Appendix A), and reflects irresistible logic and reality.

As with so many other issues in this whole area of socioscientific dispute resolution, the matter of "how" poses the most difficult question. Individual members of industry (who are the "establishment" in connection with this issue) must compete and struggle to survive. Each alone can no more take on the burden of solving the problems of the world regarding risk assessment than can the individual regarding the need for additional hospital facilities in his community. There must be joint action by those who recognize and understand their own self-interests and their societal obligations. Joint action even by individuals seeking to build a modern hospital in their own community is hard enough. Joint action by giant competitors poses far more serious issues, including complex antitrust and other legal concerns.

Fortunately there is a model to observe, the chemical industry's new Chemical Industry Institute of Toxicology (CIIT). It is too early in its life to permit full evaluation of its societal contribution, but it does represent a most significant incipient joint industry attack on some of the key underlying problems in current socioscientific disputes.

Chemical research and development have always been the lifeblood of chemical companies. Important new scientific information is highly prized and is treated with the utmost secrecy. At the same time, concern about potentially injurious side effects of valuable chemicals has grown at a startling pace. There is increasing need for more basic research into the mechanisms of toxicity and the very long-term effects of widely used compounds. We desperately require greater laboratory facilities, and more highly qualified scientists to work in them. These are fundamental societal urgencies, well beyond the capacities of individual companies to satisfy, acting alone.

In 1974, a small group of chemical companies—led effectively by M. E. Pruitt, vice-president and director of research and development of the Dow Chemical Company—met to consider what to do about these problems. No one doubted the needs, but there was great uncertainty as to *how* to proceed and, at the inception at least, even as to *whether* it was possible to proceed at all. The companies were all active and aggressive competitors, reluctant to do or participate in anything which would give another the slightest edge. The adversarial environment was sufficiently charged to pose serious doubt as to whether anyone outside the industry would consider the effort and the results produced credible. It telescopes an enormous amount of time spent in discussion, debate, negotiation, processing applications to government, and obtaining the necessary licenses and permissions, to say that these companies finally created CIIT.

CIIT is an independent scientific research and educational institution sponsored, supported, and assisted in every possible way, by its member companies. Otherwise it is totally separate and apart from them. It is devoted exclusively to enhancement of the investigation and public disclosure of information regarding the potential toxic effects of chemicals. By its charter and by laws, it is not an "action" organization in any sense. Instead, it makes its information available to anyone and everyone, in and out of industry, for each to use in whatever fashion each may consider appropriate. One important part of its work is the conduct of toxicological research regarding the important specific "commodity" chemicals, a virtually never-ending task. But more important for the long run, it is investigating the basic questions involved in the new science of toxicology, such as with respect to the causes of cancer. It is acting to improve the quality and extent of toxicology generally by seeking to develop new, faster, simpler, and less expensive testing methods, and by improving and enlarging toxicological training. It has already granted a number of postdoctoral scientific fellowships, enabling scientists to study and train with its staff. It has conducted scientific conferences and has presented, or has participated in, many other educational and training programs.

By mid-1980, CIIT's membership consisted of thirty-five leading members of the chemical industry, representing the great majority of the industry's chemical production. It had completed construction of, and had moved into, its own new $14 million research laboratory, and had recruited a staff of ninety-three. It was well on its way, with an annual budget of $8 million and growing, and sufficient objective commitment to insure that it would be a valid test of whether this kind of an effort will work.

It may well be several more years before CIIT's contribution can be fully evaluated, so that others may know whether to adopt it as a model to deal with their own problems, without being required to repeat the same kind of exhaustive investigation conducted by the CIIT founders.[31] This is because the assignments CIIT has assumed are all very long-range ones. Training toxicologists and improving the science of toxicology are tasks which take years to even begin effectively, and have no end. Inquiries into such problems as cancer-causation are extraordinarily complex and difficult, as evidenced by the intensive work already conducted without final success. Even the comparatively simpler task of evaluating the toxicity of specific compounds requires years for the study of each, and only a few can be done at a time. Certainly the kinds of scientific breakthroughs which would result in immediate acceptance and approval should not be anticipated during the near term.

Nevertheless, the auguries are favorable. CIIT research reports have already been published and circulated widely. Papers by its research scientists are more and more prominent on the programs of scientific conferences in the field, and are published in leading scientific journals. Industry, government, and even, on occasion, a labor, environmental, or consumer organization have turned to it for information and assistance.

Much remains to be seen regarding CIIT, but it does appear already to have created a place for itself as a significant toxicological resource. The toxicology of chemicals, however, is only one of a large and growing number of complex scientific problems involved in socioscientific disputes. The number will inevitably increase at an even more accelerated pace as science and technology continue their advance. CIIT is thus of greatest significance as an experiment. It tests the kind of broad, new, "establishment" approach needed to improve and enhance the communication of scientific information to the public, and to bring greater credibility to science.[32]

INTERIM SCIENTIFIC "CONSENSUS-FINDING"

The technique of science is the "scientific method." It works well for scientists in their own scientific bailiwicks. Science's troubles begin when scientists enter the entrepreneurial, competitive fray. Then many begin to employ the technique of "adversarialism," rather than that of science. The adversarial technique is foreign to the background and training of scientists. It is inconsistent with "science."

Some scientists become disgusted with the apparent need to employ the adversarial method. They withdraw from contested public-policy issues. Other scientists, in contrast, embrace adversarialism in its most extreme form. They "champion" public causes. They become expert in the use of adversarial excesses. Their employment of such a methodology, however, also evidences that they are then acting as "promoters," not as scientists. Their scientific credibility suffers, as does that of all science.

In either case—withdrawal or scientific extremism—we are all the losers. Socioscientific dispute resolution is not based on the best information that science has to furnish.

It just makes good, simple, common, ordinary "horse sense" for society to ask that scientists continue to use their scientific method wherever and whenever they purport to act as scientists, as much in socioscientific disputes as elsewhere. This chapter discusses a number of existing scientific dispute-resolution mechanisms designed to achieve this objective. It also reviews three important scientific conferences conducted during the last half of the 1970s, which suggest that a new and additional approach might also be useful. Based upon the experience gained from those conferences, the chapter proposes the scientific "consensus-finding" conference, designed to uncover scientific consensus even before all essential scientific investigation has been completed. This new kind of conference would supplement existing mechanisms by furnishing additional credible scientific input to the process of socioscientific dispute resolution.

It is a fundamental assumption of the consensus-finding approach that some of what appears to be scientific controversy in socioscientific disputes, is not really serious scientific controversy at all. A part, as discussed earlier, is essentially public-policy dispute. That is for the public as a whole to decide, not scientists alone.[1] A good part of the balance is the result of speculation, gossip, "knee-jerk" reaction and the like. These persist because the scientists involved sit isolated and alone, or with sympathizers, fed by their own perceptions and the biases of others. Ordinarily, they do not have an opportunity to trade ideas and information, and to cross-question each other directly on issues which are not yet ripe for the traditional scientific consensus. It is only when scientists get together in the atmosphere of true science that they adequately examine these preconceptions. Then, extremes become attenuated in the presence of peers, and soon they find substantial areas of agreement. The public is entitled to have the benefit of such agreements. The consensus-finding approach seeks to accomplish this for the public.

The Scientific "Consensus"

The lay person commonly defines scientific "consensus" to signify unanimity, and that is certainly close enough for most lay purposes. It is the meaning I have intended in employing the term earlier in this book. But it is not technically correct. The difference can be important when we are dealing with socioscientific disputes in which only a small part of the needed scientific evidence is yet available.

Supreme Court Justice Stewart wrote that perhaps he could not intelligently define hard-core pornography for criminal prosecution purposes, "But I know it when I see it."[2] Much the same is true of the scientific "consensus." Scientists do not seem able to define it precisely. It is not any fixed proportion or percentage of scientific opinion, although, of course, real unanimity usually means that a consensus exists. Dictionary definitions such as "harmony," "group solidarity in sentiment and belief," "general agreement," and "collective opinion" are helpful only in giving flavor to the word.

A scientific consensus can be independent of the underlying opinion to which the consensus relates. Thus, scientists may agree that there is a consensus regarding a scientific opinion, even though some take exception to that opinion (after all, it was once the consensus that the earth was flat and that man sprang from Adam). Similarly, they may be in complete agreement about the underlying opinion and yet conclude that there is *not* yet a scientific consensus about it (for example, because of inadequate scientific participation, or insufficient time for review). Scientific consensus is therefore very different from scientific opinion. Moreover, and perhaps surprisingly, there will far more often be complete unanimity regarding the existence or lack of consensus, than there is with regard to the underlying opinion. Where consensus exists, it is enormously credible and persuasive. Whatever consensus is, scientists alone known it when they see it. Accordingly, any scientific consensus-finding mechanism should itself determine the extent to which it has found consensus and then communicate that fact to the public.

Perhaps sometimes there will be scientific consensus with regard to all the scientific issues involved in a socioscientific dispute, but that will be extremely rare. So many questions will be so intimately tied to "value" issues, that separation out of the strictly scientific ingredients is virtually impossible. But scientific consensus is not significant only with respect to "ultimate" issues. It can be almost equally important with regard to subsidiary and even more remote matters, such as what kinds of, and how many, species should be included in a laboratory experiment, or the economic effects of distress surplus on pricing in a competitive marketplace. Indeed, where a dispute involves the "soft" sciences, ordinarily the only types of questions on which consensus will be possible are those involving methodology, practices, standards, qualifications, probabilities, or responsible options and alternatives open to consideration. Consensus even in such limited areas, however, helps to remove unnecessary controversy from the public dispute process.

Traditional Scientific Method

The traditional scientific method has served society well. It seems generally agreed that it has led to increased material wealth, improved health and longevitity, enhanced communication and education, and brought society to the threshold of an exciting world—which we can enjoy if only we can learn how to manage our resources. The trouble with trying to use it in the resolution of many present socioscientific disputes is that it simply takes too long to arrive at its final product, the traditional consensus. Society cannot wait that long.

Scientific data collection, observation, investigation, and research are the first steps in the traditional process. It can take years just to complete initial empirical analyses, epidemiological studies, and laboratory experiments. When finished, the original investigator prepares his conclusions, presents them to his scientific colleagues, and publishes his work in a scientific journal. Then others comment, criticize, try to repeat the work, present their own evaluations to their colleagues, and publish. The original investigator, and others, may then go back, reexamine earlier conclusions, revise, re-present, republish, and so on and on until everyone is satisfied. Moreover, one of the essential cornerstones of science is freedom, and much of this work is voluntary, taking place to suit the convenience of the scientists involved. No one sits in charge, directing the entire scientific community to come up with an answer. Researchers in academic laboratories, and foreign colleagues operating in entirely different social and economic systems—with totally different priorities—are entitled to as much opportunity to participate, and to critique, as those in industry and government. Work must be adjusted to teaching schedules, limited academic budgets, inadequate facilities, and sometimes even delay resulting from the teaching sabbatical of a participant.

Apart from actual fraud, there is little more serious criticism of a scientist than that his judgment was based on incomplete data, inadequate evaluation, speculation, surmise, or guesswork. The traditional final scientific consensus accordingly comes only after every responsible scientist has had a fair chance to participate in the investigation and evaluation processes, and only after all the relevant data, research results, opinions, and critiques, have been assembled, considered, and judged by the responsible peer group.

On occasion, this traditional approach can work fairly quickly. Nevertheless, it should be apparent why the process does not work to resolve the scientific issues in a socioscientific dispute. There we must have an answer at once. There is no way of waiting without deciding, or stopping the clock. We either do or do not continue with nuclear power. Even the "yes, but" decision, such as permitting cigarette smoking or saccharin consumption to continue provided a warning is included on the label, represents a decision. Such judgments must be based on whatever evaluation of the scientific merits is possible, even if it cannot be final for many years.

Forming the best possible public policy decisions in socioscientific disputes requires a very different kind of scientific consensus than that of the past. We need a consensus with regard to the admittedly inadequate, incomplete, and unsatisfactory scientific information at hand. We must ask our scientific community to do something which it has always been reluctant to do—to furnish us its combined best judgment right now, even though it knows that more tests, more critiques, more work are needed, and even though it may well turn out that the present best judgment is quite wrong.[3] As the final report of the June 1979 scientific "dispute-resolution" conference stated:

> Traditional science seeks to avoid conclusive decisions until all the necessary data, research results, and opinions have been assembled, evaluated, and judged so that a scientific "consensus" can be reached.
>
> We no longer have the luxury of awaiting a final scientific consensus in this traditional sense. Decisions *must* be made now. There is no other alternative. We either do or do not use a chemical; we either do or do not use nuclear power. To await the final, traditional scientific consensus may mean that the barn door was closed long after the animals escaped. We must find a scientific alternative. (Appendix C)

Specialized Scientific Panels

The need to factor scientific data and opinion—interim or otherwise—into the dispute-resolution process has of course been recognized for years. Where their own staff resources are inadequate, government agencies such as the Environmental Protection Administration (EPA), the Food and Drug Administration (FDA), and the Office of Technology Assessment (OTA), legislative bodies, industry groups, and many others often convene special scientific advisory committees and panels. These are made up of outstanding scientists, sometimes selected from National Academy of Sciences lists, and sometimes selected by the Academy itself. They review all the available scientific evidence, hold hearings where appropriate, and finally express conclusions and recommendations.[4]

The conclusions presented by these panels are extremely useful. I do not intend to deprecate their great value by suggesting that they do not always have the powerful public credibility and impact of the true scientific consensus. This is because they do not necessarily reflect such consensus. Too many scientists are excluded from the process by which such panels are selected. Even where the issue of broad scientist participation is considered—and most commonly it is not—there may be sound practical reasons for this exclusion. For example, there may be a legal requirement to go ahead and get something completed before a deadline expires, or inadequate funding. Where broad scientific participation in the panel's work is permitted, some who might like to be involved are not informed, or do not have the resources or the time to participate. In either case, the resulting exclusion is the same and impedes consensus.

Perhaps most significant, there is a perception on the part of some scientists that there is inherent, insidious bias in the panel-selection process and, as a result, in the panel's studies.* This view argues that the person appointing the panel, subconsciously (or perhaps consciously—it makes no difference) chooses people whom he believes will be partial to the result he favors; others are char-

* One scientist, reviewing an earlier draft of this manuscript, wrote: "I believe that many academic people are chosen for participation on the rather tenuous grounds that they have done so previously, because of their availability, or simply their desire to do so. I would add to this possibility that some of these experts, at least, are simply echoing what is fashionable (even in science) . . ." Private letter, October 5, 1979.

acterized as "less well qualified," "unreasonable," "extremists," or "flakey," and excluded. The National Academy of Sciences and its membership are seen as strictly "establishment," controlled and composed very largely of scientists who have "achieved" in major ways within the present "repressive" economic and social system, dependent upon government and industry for funding and other support,[5] and certainly not reflective of the views of other less "well-fixed" opinion in the scientific world. Whether or not any of this is true is really almost irrelevant. The perception itself means that the general community does not always accord the status of true scientific "consensus" to the conclusion of a scientific panel studying a controversial issue. For all these reasons, the "scientific advisory committee" approach—valuable as it clearly is—is not enough.

Since September 1977, the National Institutes of Health (NIH) has been conducting an experimental "consensus-development" program. It seeks to develop a broader base than that offered by the traditional scientific advisory panel. The public-health effects of any drug, device, or medical or surgical procedure may be evaluated. A panel of concerned persons, lay as well as professional, seeks to find consensus at an open conference. The participants include ethicists, consumers, and lawyers, as well as physicians and scientists engaged in relevant research. Expert opinion is presented, and comments are received from an audience.

Twenty conferences were conducted during the first two years of the program. The subjects considered included X ray breast-cancer screening, the use of amniocentesis, the distribution of insect-sting kits, the appropriate use of tonsillectomy and adenoidectomy, the need for radical mastectomy, and the use of estrogens to treat the symptoms of menopause. In the last, for example, the consensus panel recommended that a woman considering estrogen therapy should be clearly told of its benefits and risks, and then allowed to decide for herself what she wishes to do. Seymour M. Perry, who heads the NIH program, reported after the first two years that there had been little criticism of the program by doctors or professional organizations. Many of those who commented on it "have described it as a good thing, something long overdue."[6]

The NIH program does not pretend to limit its active participants to scientists from the pertinent disciplines, or its concerns to "strict science." By including lawyers, ethicists, consumer representatives and others, it makes clear its purpose to find a broader kind of public policy consensus, not scientific consensus alone. It is thus a kind of hybrid dispute-resolution mechanism, similar in some respects to the National Coal Policy Project (NCPP), to be discussed later in this chapter. Nevertheless, it is also relevant here as illustrating the enhanced public impact which may result if the base of participation in consensus-finding is broadened to include as many potential antagonists as possible.

Another long-standing approach to bringing better science into scientific dispute resolution, has been to improve the expertise of the decision maker. Over many years, there has been increasing delegation of complex scientific issues away from our judicial courts and into specialized bodies, such as patent courts, tax courts, and a great number of courts within the administrative agencies, such as EPA, FCC, FTC, ICC, NRC, and many others. Judicial courts will almost always defer to these expert bodies when reviewing matters decided by such bodies within the ambits of their special competences. In addition, many judicial courts have themselves developed procedures to assign complex matters to judges who may have expertise in the relevant issues. Such practices are clearly on the increase. There has been some recent objection to what a number of persons consider an improper abdication of judicial responsibility to regulatory agencies but, in general, all of these efforts at improving arbiter expertise have been useful and helpful.

Some scientists and others concerned with the lack of judicial specialization and expertise have been proposing the creation of a new "science court." That body would be charged with a responsibility to render judgments on complex scientific issues. It has been described differently by various proponents, and there is no need here to consider all of its details.[7] One aspect is critical, however. An essential feature of all descriptions of the proposed court, is the inclusion in its procedures of a substantial number of the existing elements of our adversary system, including adversary parties, discovery, cross-examination, and the like. A version of the proposal is being tested by the Engineering Society of Detroit, with considerable official support, and, of course, we should await its outcome for better evaluation.[8] At this point, however, I must express reservations regarding the adversarial scientific procedures which are central to the proposal. These move in precisely the opposite direction of that which is urged by one key theme of this book. I believe we must develop new ways to solve scientific problems in scientific ways, using the scientific method. The science court, in contrast, approves, gives structure to, and even formalizes, "adversary science." I am most hesitant to lend any encouragement to the unfortunate trend in that direction, which is already widespread enough. Moreover, I do not consider that the science court would contribute much to the improvement of socioscientific dispute resolution. It does attack the problem of inadequate arbiter expertise it is true, but that is already being dealt with in many quite acceptable ways. Lack of expertise is not the most serious part of our dispute-resolution problem. Instead, the critical need today is for public confidence and credibility. The addition of another "court" is hardly a "patch" to a dispute-resolution program which is as inherently defective as that which we employ today.

A third approach to improved socioscientific dispute resolution has been to increase the legislative involvement in such problems. As discussed in chapter 1, these disputes are essentially "legislative" in character. They involve the delicate balancing of a great number of conflicting quality-of-life pluses and minuses, goods and bads, go-aheads and stops, in the interests of society at large. One of the reasons we are having such trouble, is because legislatures have in effect "bucked" decisions of these kinds to administrators, agencies, and courts which are not designed to be representatives of the people. When we return them to the legislative arena, it is contended we place them where they belonged in the first place. There are many proposals for doing just this, such as with regard to the matter of nuclear power plant siting.

These proposals do have merit where very broad societal considerations are involved. Legislatures do make such decisions regularly. Congress has already pulled back previously delegated socioscientific decision-making when the public clamor got loud enough. This was done, for example, in the trans-Alaska pipeline dispute, where Congress decided that the energy benefits outweighed the potential environmental hazards, and in the saccharin dispute, where it concluded that people should be free to make up their own minds, at least until scientific research was further advanced. The public appears to have been satisfied with these decisions.[9]

For most other socioscientific disputes, however, the legislative process may well pose even more difficult problems and concerns than the administrative or the judicial. The conflicting political pressures are inevitably very great. Legislators already deal with an enormous number of matters. They have neither the staffs nor the time to become sufficiently well informed regarding the great number of complex issues which are involved in these matters.

Moreover, although current socioscientific disputes may appear numerous and difficult enough, in historical perspective we are probably just at the very beginning. As society matures, we will have more and more of these problems, and they will reach further and further down into the lowest levels of our society. Legislatures should only be asked to take care of the "really big ones." Congress today, for example, is struggling with the key issues of inflation, energy, and the second Strategic Arms Limitation Treaty (SALT II), and related defense-planning and budgeting concerns. Any reader examining his own state or local legislature, will find similar major issues which eclipse even important socioscientific disputes. If we can, we should develop mechanisms that work independent of legislatures, and that operate as well at the lowest and smallest levels of society as they do at the top.

This section has discussed a number of approaches to improving the manner in which scientific data and opinion are factored into the socioscientific dispute-resolution process. Undoubtedly analysis and study will suggest others. I would like to propose one—the scientific "consensus-finding" conference. This brings together, in one place and at one time, a fair representation of scientific opinion on the issues in controversy. Those scientists who wish to be involved can be involved. It is conducted openly, with a fair opportunity for public participation and observation. It employs the scientific method to produce a powerful and persuasive scientific consensus, which can serve as input into the required public-policy decision. In some cases, the scientific consensus should be dispositive.

No true scientific consensus-finding conference, as such, has yet been held. Three scientific conferences have been conducted, however, which combine to serve as my model for the proposal. Two were conducted during the course of specific, ongoing litigations relating to the herbicide, 2, 4, 5-T. They concentrated largely on chemistry and medicine, although they also included consideration of economics and other "soft" sciences. The third conference was not held in any specific litigation context. It dealt with problems relating to the use of coal as an energy resource. It dealt primarily with economics, sociology, and the soft sciences, although it also included the applied sciences involved in energy and coal production and distribution. Some discussion of each of these conferences is appropriate before turning to the consensus-finding proposal itself.

The First 2,4,5-T ("Agent Orange") Conference

2,4,5-trichlorophenoxyacetic acid, commonly called "2,4,5-T," is a member of a group of chemical compounds known as the "phenoxy herbicides." These are used in the management of crops, forests, rangelands, aquatic habitats, and rights-of-way. They function by changing the natural growth processes of plants. They are attractive to their users, because of their excellent selectivity in killing only broad-leaf vegetation while allowing the various grasses, crops, and desirable vegetation to grow more abundantly.

2,4,5-T was first introduced in 1948. It soon proved to be especially useful as a control for trees and brush; for weeds which threaten the growth of rice, conifer seedlings, and rangeland grass; and for weeds which create hazardous conditions on road shoulders and utility rights-of-way. For more than a decade and a half it stirred little concern. For example, coupled with another herbicide, 2,4-D, it was mentioned just once, and then only in passing in Rachel Carson's *Silent Spring*, published in 1962, with the comment, "Whether or not these are actually toxic is a matter of controversy."[10]

Two chance circumstances, one not really related to toxicity and the second a mistake, set off a controversy about 2,4,5-T which has raged for over a decade, and which continues unabated to this day. The first was 2,4,5-T's use, along with 2,4-D, in a defoliant known as "Agent Orange," used in Vietnam to destroy the tree cover which sheltered the Vietcong. Towards the end of the 1960s, rumors and reports from Vietnam charged that Agent Orange posed serious health hazards to the general populace of the sprayed areas. These were picked up by Vietnam protesters. Together with many other things, they became a part of the bitterness and acrimony which gave rise to the student riots, the Kent State killings, the charges about napalm, and so much of the other turmoil that characterized our later involvement in Vietnam.

The charges about Agent Orange also led to accelerated scientific laboratory research on 2,4,5-T. One study seemed to confirm the Vietnam information, indicating that the compound caused birth deformities in mice and rats, and therefore was quite possibly a "teratogen" in humans as well. This apparent scientific confirmation furnished the spark needed to begin the virtually unprecedented fight which seems destined to continue for several years to come. The fact that it very shortly turned out that the apparently confirming study was based on a mistake only added fuel to the fire. Instead of having been a study of 2,4,5-T as commercially used in this country, the test material had been contaminated with a recognized extremely toxic compound called 2,3,7,8-tetrachlorodibenzo-para-dioxin, commonly termed "TCDD" or, not quite correctly (because there are other different members of the dioxin family), "dioxin." Some amount of TCDD is an apparently unavoidable concomitant of the 2,4,5-T production process. Because of known toxicity, this is today held to well below one-tenth part of TCDD per million parts of 2,4,5-T (it was 1 ppm in 1970) in the commercially used product in this country. Somehow the study test material contained 27 parts of TCDD per million parts of 2,4,5-T—27 times as much then, and 270 times as much as 2,4,5-T manufactured in the United States today.

Instead of stopping the 2,4,5-T controversy until new research could be completed, however, the error led only to expansion of the struggle to include TCDD as well as pure 2,4,5-T. Soon other products containing TCDD were involved in the fight as well.

It would take a book to describe the ongoing struggle about 2,4,5-T, TCDD and dioxin. In fact, two books have already been written, plus a huge number of reports and articles in lay and scientific publications.[11] What is relevant here is only that the battle has included executive and administrative orders, debates, meetings, lawsuits, appeals, hearings, trials, and virtually every other kind of dispute-resolution mechanism our adversarial society has to offer, including legislative on the federal, state, and local levels. At one time or another, each of the compounds has been charged to be a carcinogen (causes cancer), a mutagen (causes mutations in subsequent generations), a teratogen (causes birth defects) and otherwise hazardous to humans and the environment. The key present strictly legal proceedings, are an administrative action by EPA seeking to end use of the product—initially provoked by allegedly recently discovered evidence of 2,4,5-T causing birth defects in Oregon; a class action by Vietnam veterans, charging that their exposures to Agent Orange have caused cancer; and a cancer-death action in Arizona, also claiming that 2,4,5-T caused cancer. Every so often another front-page newspaper article or a national television "special" appears, and adds further fuel to the controversy.[12]

This extraordinary level of public interest and concern has been quite naturally matched by an equally extraordinary level of scientific investigation. Final answers are rarely possible in science. As one finds satisfactory, or at least acceptable, answers at the ppm and ppb levels, the analytical chemists develop techniques for measuring at the ppt level and below—where they are now—and the process begins all over. But most of the scientific researchers who have worked on 2,4,5-T believe that more was, and is, known about 2,4,5-T and TCDD than virtually any other chemical compound. The critical issue in the past and present socioscientific 2,4,5-T litigations is how to factor all this scientific information into the dispute-resolution process.

A scientific conference conducted in early 1974 as a step in the process of preparing for a forthcoming 2,4,5-T hearing turned out, quite serendipitously, to be a "consensus-finding" conference as well. Several years before, in 1970, the EPA had begun a formal administrative litigation, seeking to prevent certain continued uses of 2,4,5-T. Interventions, motions, arguments, hearings, injunctions, appeals, discovery of documents and contentions, and a huge variety of other prehearing steps in the case, brought it, by the late winter of 1973–74, to the eve of final trial on the merits. By this time, the products involved and the charges had been expanded. On one side, arrayed against further use of the products, were EPA and EDF, the leading environmental action organization working in this area. On the other side, as proponents of continued use, were the U.S. Department of Agriculture (USDA); manufacturers of the product, led by the Dow Chemical Company; user groups, including the American Farm Bureau Federation (AFBF); and a number of others.

The public has an impression that trial attorneys fight only on the court "stage," and that otherwise they are buddies—eating and drinking together, and enjoying each other's company. This is sometimes true, but not nearly so often as people think. Good trial attorneys must be good actors, and good actors must "get into" and really believe their roles. In public interest cases especially, attorneys often become just as much "activist" as their clients. Then the bitterness, venom, and even hatred which can characterize a court fight become most vicious. The attorneys can get so involved, that they lose all sense of perspective, are unable to conduct themselves sensibly, and even commit crimes.[13] Sometimes an experienced attorney will deliberately taunt his adversary and "drive him up the wall," just to produce such an error, or to demonstrate to the judge and jury that he and his client are not the unbiased, impartial, aggrieved persons they claim to be.[14] Some of this happened in the early 2,4,5-T proceeding, so that the litigation atmosphere between some of the attorneys was often charged with far more than the usual antagonism. On occasion, even the social amenities were absent, with one attorney refusing the common legal courtesies to another without a direct order from the Administrative Law Judge.

In this atmosphere, one side determined to hold a "rule-of-reason" conference as a part of its final trial preparation. It is to be emphasized that this was not designed as a new dispute-resolution mechanism at all. Rather, it appeared that playing the litigation "game" in the usual way, by holding evidence "close to the vest" so as to be able to "spring" it in surprise on cross-examination or at another opportune moment, could well be self-defeating. There were so many witnesses involved, and the scientific evidence was so complex, multifaceted, and far-reaching, that serious errors might result. Scientific experts need a fair chance to consider the comments and critiques of colleagues and scientists in other disciplines. Moreover, following a full, open debate and disclosure of all the relevant scientific information, each scientist would undoubtedly feel more comfortable in expressing his final conclusion, or in endorsing the conclusions of others. As stated earlier, many scientists are not happy with the cross-examination, "sand bagging" techniques of our litigation system. They require even more careful preliminary reassurance than lay witnesses, so that what finally takes place at the trial itself does not involve surprise and the need to express a view which is not "scientific," because it has not been adequately reflected upon and considered. Finally, the full "rule-of-reason" examination and disclosure of all the scientific evidence might well suggest modifications of the proponents' litigation approach. New understanding or information might suggest that more testing and research were required, or that some preconceived view of the scientific evidence should be modified. No one suggested, or probably even dreamed, that the "rule-of-reason" conference would in fact end the litigation.

Under the leadership of Fred H. Tschirley, a noted plant biologist who was then USDA Pesticides coordinator, an "open" scientific conference was convened in Washington, D.C., on March 8 and 9, 1974. The conference was not designed to seek consensus, but to review all the pertinent scientific information available regarding 2,4,5-T risk and benefit, in order to assure that the best possible evidence was available, with no mistakes resulting from inadequate discussion and consideration. All those who were expected to appear as proponent scientific witnesses were of course invited, plus members of the independent scientific community who had background and information on the issues, and who thus might be able to assist in the learning process. Anticipated "adversary" scientists from EDF and EPA were not initially invited. However, the conference was made known to the Administrative Law Judge, and to the other parties, and publicized as widely as possible in the 2,4,5-T scientific community. EDF/EPA requested an invitation to attend, which was granted, and three adversary scientists attended, two from EDF and one from EPA.

Ninety persons participated in the two-day conference, including twenty-eight scientists from academe and other government agencies not involved directly in the dispute. Scientists came from such respected institutions as the National Academy of Sciences (NAS), the National Center for Toxicological Research (NCTR), the National Institute of Environmental Health Sciences (NIEHS), Purdue, Brown, Duke, Carnegie-Mellon, Vanderbilt, and California Institute of Technology. There were plenary sessions; four separate toxicology "workshops" in teratology, human toxicology, "other" toxicology, and carcinogenicity/mutagenicity; five chemistry workshops in environmental impact, analytical methods, residues, sources of dioxin, and statistics; and, finally, a "rule-of-reason" session which explored the methodology involved in risk/benefit analysis. Rapporteurs were present at each of the sessions to summarize the discussions and conclusions both for the plenary and for the final overall report that was to be submitted to the Administrative Law Judge, and to be circulated widely (including of course to adversary parties). Adversary scientists not only had access to all and said what they had to say, but one of them "cross-questioned" and—perhaps to the surprise of a few proponents—contributed to the value of the exchanges.

The conference had three important results. First, those participating became more and more comfortable with their scientific views. The conference ended on a note of great scientific enthusiasm, with proponents and their scientists fully satisfied with their presentation and ready to go into the final phase of the litigation battle. As a strictly adversarial, trial-preparation tool, the "rule-of-reason" conference had clearly proved itself.

Second, the uniqueness and importance of the meeting, and its deliberations and conclusions, "got through" to EPA. Senior EPA executives, ordinarily far too busy with other matters to get intimately involved in the merits of a litigation turned over to the legal staff for handling, began to get information which was not filtered through or "colored" by the adversarial perceptions of their trial counsel. A senior-level, fresh examination of all the scientific evidence was undertaken, conducted by the client-principal and its scientists, and not by its lawyer-representatives. The "rule-of-reason" conference thereby achieved a second, albeit unanticipated, important result. It permitted client to communicate directly with opposing client, instead of through adversary counsel as usual.

Such direct communication between clients represented by counsel can be most useful (assuming, of course, that counsel are kept informed and their advice is considered throughout.) This is because clients frequently perceive things very differently than their attorneys. Thus, for example, the attorney receives a legal fee, while the client pays it. Litigation is a main business of the trial lawyer, while it is a diversion from the main business of the usual client. When clients communicate directly, they deal with their own perceptions and can sometimes resolve issues which might otherwise involve confrontation.[15]

Although clients have a clear right to deal directly with each other, such communication is unfortunately all too rare. Partly this is because clients do not always appreciate the value of dealing directly under certain circumstances.[16] More generally, however, it is because the trial attorney is subject to an ethical prohibition against speaking with his adversary's client in the absence of the adversary, or of specific permission. This prohibition becomes misconstrued as prohibiting direct client discussions as well. The "rule-of-reason" conference cut right through all this.

Lastly, and most important insofar as the "consensus-finding" proposal is concerned, the scientists who attended the 2,4,5-T conference found an enormous amount of agreement regarding matters which had been thought to be the subject of dispute. This was the first time that a diverse scientific group of this kind, including adversaries, had been brought together to examine the opposing evidence and contentions with regard to 2,4,5-T. Until then, most had operated alone or in small groups with similar views, examining bits and pieces of information which made it impossible to piece together the full picture. Moreover, much of what they received was filtered through legal counsel, who added their own adversarial perceptions. As the conference analyses, discussions, and debates continued, however, it became more and more clear that many apparent scientific differences were not differences at all, or were really differences in the kinds of risks people believed worth taking—value differences, and not scientific differences. The areas of agreement were far greater than the areas of difference. Even as to the latter, there was agreement that what was needed was more investigation, and that no scientific debate was in order.

The final report of the 2,4,5-T conference, prepared by USDA attorneys and filed in the EPA's administrative proceeding, is available publicly.[17] Whatever "public-policy" conclusion one draws from it, it quite clearly demonstrates that, as of the time of its preparation, there were no really important scientific questions which required presentation in a full-blown adversarial hearing.

On June 24, 1974, after concluding EPA's scientific review of all the evidence, Deputy EPA Administrator John Quarles called the parties together and announced that EPA was terminating its cancellation proceeding. William Upholt, EPA's then senior science advisor on the relevant issues, presented a history and assessment of the dispute and then concluded:

> The conclusion seems clear that there is insufficient evidence regarding residues of 2,4,5-T, and the dioxins in the environment now, and also on the hazards associated with these residues from the remaining registrations of 2,4,5-T pesticides.
>
> On the other hand, we must intensify efforts toward fuller analysis of possible hazards from TCDD and other dioxins from all trichlorophenol-derived pesticides as rapidly as improved analytic procedures and properly designed testing permits.
>
> The results of such analyses may or may not justify reopening the question of hazards of 2,4,5-T as well as the related pesticides at a later time.[18]

Whatever future scientific research and investigation might suggest, EPA's "public-policy" decision on June 24, 1974, was that the benefits of permitting continued use of 2,4,5-T, outbalanced the hazards. People might differ with this value judgment; no one differed sufficiently with the scientific evaluation to complain legally. As the result of the rule-of-reason conference, "science" had thus been factored into "public policy" with enough credibility to at least end the legal fight for the time being.

The Second 2,4,5-T Conference

Unlike old soldiers, socioscientific disputes do not always "fade away." For a few years after's EPA's 1974 withdrawal, government administrative proceedings remained quiescent. Not everyone was satisfied with EPA's public-policy decision, however. Other fights continued and the pressures on government persisted. By late 1978, preliminary EPA administrative action had once again been initiated, and by mid-1979, this had ripened into full-blown adversarial judicial and administrative proceedings between essentially the same parties.

Many scientists are unhappy with the adversarial fashion in which we seek to resolve important scientific questions. Fred Tschirley is one of them. His interest in scientific dispute resolution continued after he left government and became chairman of Michigan State University's Department of Botany and Plant Pathology. He believed that the 1974 2,4,5-T conference, which he had convened and chaired, contained ideas upon the basis of which a new approach to scientific dispute resolution might be developed. Accordingly, when the 2,4,5-T dispute grew reheated and the American Farm Bureau Federation (AFBF) approached him in 1979 to chair another conference, he accepted the invitation. He selected another scientist, Theodore C. Byerly, to serve with him as conference chairperson and convener. Tschirley, Byerly, and William Upholt, who had headed the EPA's 1974 scientific review, acted as conference overviewers.

Unlike the 1974 conference, the 1979 2,4,5-T conference was labeled from the start as a "dispute-resolution" conference. It was Tschirley's hope that it might serve as a model for other conferences to follow. With this in mind, he sought the broadest possible participation of scientists with opinions in all parts of the 2,4,5-T spectrum, and wide public "observer" participation. Invitations were extended personally to all scientists known to have significant knowledge or information bearing on 2,4,5-T, in government, academe, industry, and environmental and consumer organizations. Unfortunately, not all invitees could or would attend. As gaps in participation began to appear, Tschirley sent follow-up invitations to organizations with known positions, requesting that they identify scientists who might be invited. Relations with the media were established through the AFBF public-information framework; opinion leaders from the lay public were invited to participate as observers, and were furnished an opportunity to comment and criticize. The conference was organized within a very tight time schedule so that its conclusions could be factored into an EPA administrative review scheduled for the late spring of 1979. Except for the resulting constraints of time, little effort was spared to encourage the broadest possible scientific and public involvement in the conference.

The 1979 conference was held in the Washington, D.C., area, on June 4 through June 6. It consisted of plenary sessions and workshops similar to those which had been held in 1974. Forty-nine scientists attended from twenty states in this country, plus eleven from New Zealand, Switzerland, Italy, Germany, France, Sweden, and Canada. Fifty-one additional persons, making a total of 111 in all, attended as observers, including one representative of the USSR. For three days the scientists reviewed all the scientific evidence and contentions, discussing, presenting, debating, acknowledging, and, finally, once again reaching consensus on an enormous number of the substantive scientific issues which had appeared to be in controversy. Independently, and of far greater importance to the consensus-finding proposal, the conference drew important procedural conclusions as well. The group was convinced that its "pilot effort in developing an effective dispute resolution mechanism represents a contribution to the process by which our society may come to judgments with regard to scientific matters." The nontechnical portions of the conference's proceedings are reproduced in Appendix C. The concluding paragraphs are persuasive:

> We believe that this conference has pointed the way towards a new dispute resolution mechanism which will help society to resolve many of the complex problems with which it is faced.
>
> We are aware of our scientific limitations. We are primarily experts in biology, chemistry, statistics, and economics. Other disciplines must be involved in the fashioning of any long-range dispute resolution mechanism of more general application. Not only would this include scientists from other "hard" disciplines, such as nuclear physics and computer technology, but from the "soft" sciences, including sociology, law, government, political science and the like.
>
> The Conference has concluded. We call upon those who observe and approve, to convene a further conference to deal with the socio-scientific dispute resolution mechanism as such, rather than with any specific controversial issue.
>
> Science is neither good nor bad. It is how science is used by society that counts. We believe that science can help make this world a better place in which to live. We hope that society will learn how to employ science to its benefit. We as scientists will contribute our valiant effort to that end. We hope that others will join with us in the pursuit of the goal of a better life for all.[19]

I have a deep admiration for Fred Tschirley and his colleagues and for Rick Main and his American Farm Bureau Federation associates who struggled so mightily to pioneer in this most difficult, multidisciplinary area. The product they produced is an important one. It is not critical to suggest—with hindsight—ways in which even better results might be achieved at another conference.

The impact of the conclusions reached at the 1979 conference has been significant in a number of scientific respects. Unfortunately, however, the public impact, and the impact on pending controversies, was not as great as the 1974 conference. This is because there was very little public comment about the conference in either the scientific, the legal, or the trade press. What little was published was not generally supportive either of the substantive scientific results of the conference, or of its experiment in dispute resolution. Those administering the conference did their best to encourage media coverage and reporting. They were unsuccessful. A few who were suspicious of the conference's purpose and approach also voiced their objections. In contrast, they were able to get through to the press with effect well beyond their apparent numbers.

Before the conference was held, *Science* magazine reported:

> A scientist at the National Wildlife Federation . . . says that Tschirley has bent over backward to obtain a balanced attendance, and that he "has no quarrel with the way it's being set up so far."
>
> Officials of the Environmental Defense Fund, which is bringing suit to support an EPA ban of 2,4,5-T, have a different view of the conference, and are urging environmentalists not to touch it with 10-foot leafy trees.[20]

Chemical and Engineering News, published by the American Chemical Society, reported following the conference:

> Most of the more prominent scientists who oppose 2,4,5-T use were not at the conference. And EPA scientists, who were intended by the organizers to counterbalance scientists from Dow in the various workshops, also did not attend.
>
> The result was a conference heavily weighted by scientists in favor of using 2,4,5-T. Eighteen of the participants and observers, for example, were Dow employees.
>
> Robert Harris of the Environmental Defense Fund, one of the opposing scientists not at the conference, . . . blames short notice for the lack of participation by opposing scientists. . . . Harris is also somewhat suspicious of the conference because its sponsor, the American Farm Bureau, has sued EPA over the emergency suspension of 2,4,5-T. "It didn't seem to me that this conference was organized for a value-free discussion of scientific data," he says.[21]

Pesticide & Toxic Chemical News, an industry trade journal, reported:

> The meeting . . . appeared to be influenced, if not dominated, by representatives of Dow Chemical Co. There were nine Dow scientists among the 56 participants in six "workshops" and nine more attended as "observers."
>
> Spokesmen for AFBF deplored the virtual lack of representation from public interest environmental groups, which had been asked to participate. EPA, invited to furnish a participant on each panel, was represented by only three observers, who were not conspicuous. . . .
>
> Corridor gossip during the meeting included an unverified report that Friends of the Earth had suggested a total boycott of the Farm Bureau meeting by all of the hardline environmental organizations.[22]

Most revealing as to the reaction of the media is what was published about 2,4,5-T during and after the period of conference planning, compared with the silence about the conference. In April 1979, the *National Law Journal* ran a front-page feature story about the Vietnam veterans' Agent Orange class action and dioxin; not a word about the scheduled conference (of which it had been informed).[23] On May 27, 28, and 29, less than a week before the conference began, the *New York Times* ran a lengthy, three-part, front-page series on Agent Orange. The writer covered so many aspects of the matter so thoroughly as to suggest that he had been given a word-space assignment and was sometimes looking for "fill." But not a word was included about the scheduled conference.[24]

On July 10, the *Wall Street Journal* ran a front-page feature piece on the Seveso, Italy, dioxin episode, which had been prominent on the conference agenda, and regarding which several of the Italian scientists participated, but not a word about the conference, which had reached critical conclusions regarding Seveso only a month before.[25] On July 11, the *New York Times* even devoted two six-and-a-half-inch columns of space ("Long Distance Swimmer Foiled") to the trivial story of a swimmer who had *failed* in his plan to swim around Manhattan Island to call attention to the plight of the Vietnam veterans exposed to Agent Orange; again, there was not a word about the conference. The inference seems inescapable from all this that the media, and therefore the public, viewed the conference program with doubt and suspicion, and probably as little more than a step in an adversarial litigation process.

The comment of one of the EDF counsel with regard to the 1974 2,4,5-T conference, casts further light on the manner in which some persons, including the press, viewed the 1979 conference. A lengthy 1977 *New Yorker* magazine article devoted to 2,4,5-T discussed the first conference, and reported as follows:

> About a month before the E.P.A. hearing was to be held, "Dow and the Department of Agriculture put together what I can only call a sham conference on 2,4,5-T," [the EDF attorney] continued. "It was supposed to bring together the biggest experts on 2,4,5-T and produce the definitive word on the subject. However, it turned out that the invited parties were mostly Dow employees and U.S.D.A. folk, who proceeded to give as formal papers what in essence was their forthcoming testimony at the E.P.A. hearings. We suspected that they were doing this so that if there were any conflicts in their proposed testimony these conflicts could be ironed out before the people involved went into the hearings. In other words, it was a dry run to make everybody feel comfortable—a psychological ploy by Dow's attorneys, in my opinion. The Environmental Defense Fund, although an interested party, hadn't been invited to participate. But we found out about the meeting and did manage to get the scientist who was working for us in as an observer.[26]

It seems a fair inference that EDF, and many of its environmental colleagues opposing 2,4,5-T, viewed the 1979 conference as no different from its 1974 predecessor. At least one of the conference chairpersons was the same; many of the participants were the same; the adversarial atmosphere was the same. The EPA, in contrast, may have viewed a conference billed as a "dispute-resolution" conference as one which would consider "public-policy" issues as well as strictly scientific subjects. For EPA personnel to participate in such an affair on any official basis might arguably be an abdication of EPA's responsibility under law to render the "public-policy" decision strictly on its own; discussions at such a conference might be treated as improper *ex parte* exchanges with outsiders during a dispute.

There are important conclusions to be drawn from the experience of the two 2,4,5-T conferences. The longest reasonable time frame is to be preferred, in order to permit anticipation of, and response to, potential objection and opposition; and to encourage wide public and media participation. Great emphasis must be placed on the details of organization and administration to avoid even the appearance of bias in selection or deliberation. If the conference is to include "policy" as well as "scientific" issues, full, active participation in the conference should be accorded to nonscientists who have a real interest and concern in the matters under analysis.

The compelling significance of these two conferences is that scientific consenses can be found with regard to important substantive issues being contested in ongoing litigations. These consenses were dispositive of the 1974 litigation, but not so in 1979. Why the difference? The answer to this question will suggest how the process may be changed so as to give scientific consensus greater impact in socioscientific disputes.

Neither party considered the 1974 conference as a new dispute-resolution mechanism. Even its sponsors treated it as a step in the process of preparation for hearing. Their litigation philosophy—the "rule-of-reason" approach—was new, but their adversarial objective to win was the same. Understandably, then, their litigating adversaries also treated the meeting as strictly a trial-preparation tactic and said so. They sought only to use the conference as trial preparation for themselves by obtaining invitations so that their own observers could get some free "discovery." They made no real effort to block the process by boycotting it, by well-placed media and other public comment, or otherwise. They were thus caught unaware. The effect of the scientific consensus was immediate and critical as a result. The litigation was terminated voluntarily, with a total win for the conference sponsors.

In contrast to the 1974 conference, the 1979 conference was cast from the outset and throughout as a dispute-resolution conference. This time Tschirley and his colleagues were not involved in the ongoing litigations in any way. Their purpose was to improve the process by which our society solves scientific problems. However that was not the way in which the litigating adversaries to those sponsoring the conference saw it. They had already been burned once, in 1974, and had learned their lesson about what such a conference might do to their litigation. Moreover, the "dispute-resolution" title to the conference seemed clearly to include "public-policy" as well as strictly "scientific" issues within the scope of the deliberations. Yet active conference participation was to be restricted to scientists in the relevant disciplines. Typically, environmental and consumer organizations have limited budgets, and few scientists in their employ. Thus they could not match industry scientist representation, at least in numbers, and were prevented from having others present as full conference participants even in the "policy" deliberations. Under the circumstances, it was perhaps inevitable that EDF and Friends of the Earth would react with doubt and suspicion, and treat the conference as another industry tactic designed solely to "win" the pending litigation.

The inclusion of policy issues in the conference, may well also have contributed to EPA's apparent concern that the meeting would impinge upon its public-policy, decision-making function. As a kind of snowball effect, the reservations of these groups, as evidenced by their statements and absence from the meeting, undoubtedly contributed to the suspicions of the media and its reluctance to participate. It thereby limited the effect of the conference even further.

The experience of the 2,4,5-T conferences teaches us that scientific consensus can be found, even with regard to issues involved in ongoing, bitterly contested litigations. The impact of such consensus can be dispositive of a litigation. But the consensus must be limited to strictly scientific issues to the maximum possible extent. Otherwise, adversary and arbiter may prevent consensus or limit its significance and impact. Moreover, and in any event, scientists acting as scientists should limit their claim of expertise to scientific matters. As stated earlier, they have no greater right to decide "policy" issues than anyone else.

The National Coal Policy Project

The National Coal Policy Project (NCPP) is an effort to cut through some of the environmental obstacles that stand in the way of increased industry utilization of coal. Like the 2,4,5-T conference, it also is based on the "rule of reason,"[27] and seeks scientific agreement where possible. Otherwise, however, it is very different in approach from either of the 2,4,5-T conferences.

The NCPP is a highly structured, long-range, continuing program. It has nothing to do with any litigation. Its first major conference was not conducted until more than a year after NCPP was organized. It is an experiment in a new kind of accommodation,* instead of an aspect of hearing preparation—as the first 2,4,5-T conference—or an

* The word "accommodation" is employed to describe a form of voluntary dispute resolution. It avoids what may be inapplicable connotations suggested by such traditional terms as "mediation," "conciliation," "negotiation," "settlement," or "arbitration."

effort in dispute resolution, as the second 2,4,5-T conference was billed. Agreement on scientific matters is of great importance, of course, but—unlike either of the 2,4,5-T sessions—it is quite subsidiary and almost incidental to the goal of reaching accommodation. It focuses on the "soft" sciences of economics and sociology, and on the applied sciences involved in energy and coal production and distribution, rather than on biology, chemistry, and statistics. Its entire organization and conduct are also very different, so that NCPP adds a great deal to the foundation of understanding on which a new dispute-resolution mechanism can be constructed.

The NCPP is, in one sense, "legislative" in character. Its participants commonly (not always) seek to reflect the views of their constituencies, in the hopes that those constituencies will generally endorse the conclusions reached. Participants stand entirely on their own feet, however, and have no power to speak for, or bind, the organizations of which they are members. Only persons from industry and environmental organizations are involved at this point. Following an early unsuccessful attempt to include labor and consumer groups, the decision was made not even to try to involve government, labor, consumer groups and others except as public observers. Absolute impartiality in administration is almost a fixation, so that neither industry nor environmental caucus will have any cause for complaint of bias—conscious, subconscious, or otherwise. The impartial plenary chairman, Francis X. Quinn, S.J., assistant dean of Temple University's School of Business Administration, is a well-known arbitrator and mediator. These skills have been of obvious advantage to his conduct of conference sessions. Mediation, as such, however, has been employed only on very rare occasions.

The NCPP continues to function as this is written. If analysis of its teachings is to be most useful with regard to the scientific consensus-finding proposal, however, it must concentrate on one completed "stand-alone" event in its experience. That is the February 1978 Plenary. It too was a conference seeking to help resolve an important socioscientific problem on which all the evidence is far from in.

As defined at its outset, NCPP's purpose is: "to bring together individuals from industry and environmental organizations for the purpose of achieving a consensus on a detailed plan to permit the responsible use and conservation of coal in an economic and environmentally acceptable manner."[28] Such a plan could also serve as a cornerstone of United States long-range policy, and reduce dependence upon imported oil.

The 1973-74 Arab/Middle East oil boycott brought home to the American consumer, for the first time, the seriousness of our national problem. Long gasoline lines, rising prices, inadequate fuel for winter home heating and industrial production, and a host of similar problems were personal and immediate. There was great pressure on all segments of society to take corrective action. There were demands on government especially to develop a national energy policy. This was to include conservation, use of coal, solar energy, nuclear power, water and wind power, and all other possible alternatives. When the boycott ended and, for a time, world oil production even came into surplus with price "softness," however, other priorities interfered with the adoption of a policy. Irrespective of the obvious long-term seriousness of the problem, and apart from a relatively small number of concerned persons including those who initiated the Coal Project, most of American society returned pretty much to "business as usual."

Despite its present reliance on imported oil, the United States is not really an "energy poor" nation. We have enough coal to last for several centuries at present energy-utilization rates. Coal was once a major source of energy. Cheap oil, however, replaced it to a very great extent. In the years following the demise of coal, our society came to place great value on environmental and safety concerns. Because coal was not of sufficient significance, we did not develop environmentally and economically acceptable means to be able to return to its use as our major energy source.

The problems of greatly expanding coal use are as complex and difficult as those with respect to any herbicide; many would consider them more so. They include safety in mining—always a hazardous occupation—air pollution resulting from the burning of coal, "highwalls" and other environmental damage caused by strip-mining, and an enormous number of other potential adverse effects, which, to some, are more serious than the benefits of increased use. It is not necessary here to detail or discuss these problems; suffice it to say the NCPP's first published report, entitled "Where We Agree," runs 814 printed pages, in two volumes. Yet it does not even pretend to cover all of the problems incident to expanded coal use.

The NCPP is sponsored by Georgetown University's Center for

Strategic and International Studies (CSIS). It is funded by a broad spectrum of contributors from industry, foundations, and government. It is directed by Francis X. Murray, an economist, teacher, and author, with a very broad experience in energy policy matters and no known bias except a commitment to energy problem solving. It has an "industry caucus" and an "environmental caucus," each of which has a separate chairman. The basic working groups are the task forces, each composed of an equal number of industry and environmental members, and each of which also has its own industry and environmental chairman and cochairman. These task forces obtain, analyze, and evaluate facts and data, develop alternatives, and make recommendations regarding the specific issues within their jurisdictions. There are five: Mining; Air Pollution; Conservation and Fuel Utilization; Transportation; and Energy Pricing. In addition, there is an ad hoc task force dealing with the matter of emission charges, formed when it appeared that this effort cut across the assignments of a number of the standing groups.

The work of the task forces is overseen by the Plenary Group, which is composed of the industry- and environmental-caucus chairmen, and all task force cochairmen and vice cochairmen, and chaired by Quinn. It decides on NCPP positions, by consensus. In addition, there are two other primarily administrative groups: the Technical Staff, which provides research and analytical and editorial support, and the Planning and Budget committees, which organize, administer, and provide financial review.

Following NCPP's organization, the participants and the administration labored long and hard for more than a year before the February Plenary. The industry and environmental volunteers researched, studied, met, debated, negotiated, and even attended extended field trips to get firsthand knowledge of the special problems involved. While this important substantive work was going on, Murray laid the critically important foundation for the wide public reception and impact which he appreciated would be so essential to the success of the effort. Public observers were kept informed and involved as matters progressed, and when the Plenary was scheduled, encouraged in many ways to attend. An independent public-information consultant was retained to assist in dealing with the media. Close relationships were developed with journalists and writers, so that they would be fully informed as matters progressed, and would have a personal stake in communicating outcome when the results were finally made public. Journalists were even permitted to observe the working sessions of the task forces.

As matters progressed, it began clearly to appear that NCPP was achieving substantial success in finding accommodation between groups which had been vigorous adversaries. Former bitter opponents began to come together, to develop respect for each other's views, to reexamine their own positions, and to modify and attenuate what they had earlier thought, in the interests of the broader society. A *Fortune* article written before the February Plenary described it this way: "Even the participants are astonished by their success. Probably never before have so many industrialists and environmentalists found common ground on such an array of issues. Some of the project's hundreds of recommendations—which cover matters ranging from coal mining to the pricing of electricity—are being made public this month."[29]

One of NCPP's important contributions has been its demonstration that accommodation is possible, even in major environmental disputes. Clearly this was a consequence of very careful planning, organization, and structure, all designed to avoid the most remote suggestion of bias or prejudice, or "loading" of deliberations. Laurence I. Moss and Gerald L. Decker, the first environmental and industry-caucus chairmen respectively, approached their adversarial assignments with great dignity and responsibility. They conducted themselves with the maximum possible impartiality. Murray and Quinn, who were not cast in roles as adversaries, contributed equally to the creation of the atmosphere of trust and confidence within which accommodation must breed.

The subject of environmental accommodation as such is not pertinent to the special "socioscientific dispute-resolution" subject of this book. The NCPP's teaching in this respect, however, is of great significance. It has demonstrated that environmental accommodation is a viable process, which should always be considered as an alternative to dispute resolution by confrontation. For those interested in pursuing this area, the following is a listing of some of the better-known organizations already at work on developing new approaches:

American Arbitration Association
Aspen Institute for Humanistic Studies
Conservation Foundation
Environmental Balance Association
Food Safety Council
Minnesota Public Interest Research Group
New England Energy Policy Council
Office of Environmental Mediation, University of Washington
RESOLVE—Center for Environmental Conflict Resolution
Rocky Mountain Center on Environment.[30]

Murray's extremely careful organization and planning paid off in other respects as well. The February Plenary was a major societal affair. Unlike either of the 2,4,5-T conferences, it was attended by a large number of public observers from many different segments of society, this time including government and the media, plus of course environmental groups, whose members had been participants all along. This had two dramatic and most important consequences.

First, the realization that they were in the center of the public stage, and being observed carefully, heightened the participants' consciousness of the need to consider the public interest paramount to special and parochial concern. Some who had already negotiated to what had appeared to be a final conclusion—perhaps because of fear of appearing to be weak, or of losing constituents—once again reexamined their own attitudes and opinions. Unchangeable views were changed; irreconcilable positions were reconciled. What had already been unprecedented agreement became even more so. The morning following the Plenary, the *New York Times* reported, on its front page, that the group reached agreement: "on more than 200 steps that could help the nation shift from oil to coal as a prime source of energy in ways that are both environmentally tolerable and economically sound."[31] To paraphrase Marshall McLuhan, it seems clear enough that, in some respects, the presence and interest of the media helped create the message which the NCPP participants communicated to our society.

Second, the general public impact of the Plenary was immediate and extraordinary. All around the country, front-page news stories, and radio and television, repeated the observations of the *New York*

Times. These were followed by major articles in leading national publications; appearances on widely received television programs, such as the "Today" show; invited testimony before legislative committees and appearances before important government, industry, and environmental groups; and an enormous volume of other public comment and interest. The NCPP has not solved America's coal problems, but its contributions have been significant.[32] It has relieved heat and rancor on issues in which at one time it was difficult even to exchange views; it has narrowed areas of difference and has isolated opposing contentions; where an industry versus environmentalist dispute develops, its conclusions are persuasive to a judge or other arbiter, as compelling expressions *ante litam motam*; a number of actual legislative and regulatory proposals have been adopted.

Despite NCPP's enormous success, of course there have been objections to its efforts. Some in industry, although less outspoken, appear to join in the stated reservations of environmental representatives. Those are, in substance, that the project is an elitist adversarial device. They argue that self-appointed spokespeople purport to speak for interests which they do not represent, and concerning which they have inadequate information. They believe that such problems are far better handled in legislatures, or in traditional court and agency litigations.

In comparison with the great enthusiasm with which NCPP has been received, these objections are minor. Nevertheless, it is necessary to set them forth, so as to be able to learn as much as possible from the NCPP experience. Before the Plenary, *Science* reported:

> The leader of the lobbying effort for the strip-mining bill, Louise Dunlap of the Environmental Policy Center, believes the project is at best irrelevant and at worst a device likely to be exploited by the coal industry now as it mobilizes to weaken implementation of the new strip-mining legislation.[33]

Following the Plenary, *Business Week* reported:

> When the NCPP released its recommendations last week, the Environmental Policy Center, Washington's most influential environmental lobby, issued a statement disavowing both the process and the recom-

mendations of the project. The EPC's Louise Dunlap charged that the recommendations would weaken the strip mine law and the new Clean Air Act amendments already on the books, and that most of the concessions were made by the environmental side.[34]

One specific environmental objection was that the effort was a kind of "legislative" approach, which failed to include other important segments of society. *Science* reported:

> Some environmentalists—such as Louise Dunlap, . . . and Richard E. Ayres, a Natural Resources Defense Council attorney and lobbyist for the Clean Air Act Amendments—have regarded the NCPP as a presumptuous exercise from the outset and have refused invitations to take part in it. They believe that meaningful compromises on issues of the kind raised in the strip-mining and clean air debates are best reached through the legislative process, which is open to all affected interests and which everyone knows to be "for real." Moreover, Dunlap and Ayres think that by agreeing to allow highwalls and tall stacks under certain circumstances the environmental participants played into industry hands on two key points on which environmental lobbyists managed to prevail in Congress.

Science added its own observation that

> the project participants represented only themselves. Furthermore, some interests with a major stake in coal policy, such as labor unions and Indian tribes, had no part in the project at all.[35]

The *Business Week* article which reported Dunlap's objections above, added:

> The EPC also pointed to the absence of Indians, ranchers, miners, farmers, railroad interests, and others on the task force as evidence that the project was narrowly based. Says Dunlap: "It has been a very elitist process". . . . The Sierra Club, the nation's best known environmental group, . . . backed off recently by instructing Michael J. McCloskey, co-chairman of the NCPP's mining task force, to issue a disclaimer to the effect that he was participating as an individual, not as a club representative.

The 2,4,5-T conferences suggested what might be accomplished with a longer timetable, greater emphasis on organization and structure, and admission of nonscientists to deliberations regarding policy issues. The NCPP Plenary demonstrated that these measures will work to encourage broad participation and impact. It also confirmed the view that the inclusion of "policy" issues in conference deliberations is the single greatest source of objection.

I do not suggest that NCPP should exclude "policy" issues from its consideration. That would undoubtedly eliminate NCPP's value as a new form of environmental accommodation. On the other hand, I believe that consensus on scientific issues alone can itself be a significant aid to the resolution of most socioscientific disputes, and will avoid some serious representational problems when public policy is also included. If such consensus is to be the objective, NCPP teaches us that a major effort must be made to limit conference deliberations and conclusions to "scientific" issues to the maximum possible extent. The public will then decide policy issues through administrative agencies, courts, legislatures, elections, market power, and whatever other means are available. Our history shows that it can hold the reins where necessary.

The Scientific "Consensus-Finding" Conference

I return now to the proposal for the scientific consensus-finding conference. Its purpose is to employ scientific method to resolve interim scientific issues. Its conclusions are to be factored into the public-policy dispute-resolution mechanism—judicial, administrative, legislative, public at large, or otherwise. It will function on the local and state levels just as well as on the national and international levels.

The three precursor conferences, with regard to 2,4,5-T and coal use, were entirely different. Yet the manners in which they were conducted do serve as operational and administrative models. More important, the experiences gained from those conferences furnish certain guidelines, which are set out below. Adherence to the guidelines will maximize the potential for true consensus and the dispute-resolution impact of the conference.

"Science," Not "Policy"

The consensus-finding conference must limit itself as strictly as it can to exclusively scientific issues. Unlike 2,4,5-T and NCPP, it must make a major, conscious effort to avoid conclusions as to "policy" and "value." Here scientists have no proper claims to special expertise. Where confusion of "science" and "value" is unavoidable, as will often inevitably be the case, the conference must identify the areas in which it speaks in "quality-of-life" terms. It must disown special expertise or authority on such matters.

There have always been some who believed that the learned and the skilled should rule society, but that is not our way.[36] It seems clear that many of those who objected to past conferences did so because they were unwilling to give up even a modicum of value decision-making to experts. Certainly government representatives cannot be expected to participate where their own dispute-resolution obligation, imposed by law, even appears to be assigned to, or compromised by, some self-appointed citizen group.

Limiting the conference to scientific issues to the maximum extent will also encourage the widest possible participation. It will avoid the damning "elitist" label, identifying those who arrogate the public-policy decision-making power.

In order to achieve its scientific purpose, the conference should generally avoid "ultimate" decisions, such as whether a plant should be licensed, whether a product should be distributed, or whether an organization has violated the law. In dealing with an antitrust problem, for example, it would limit its deliberations to identifying the effects of various actions, describing the relevant history, outlining reasonable options, specifying the lead time required to achieve certain goals, summarizing business practices and the like. It would not purport to find consensus that there is a monopoly, or that a practice is unlawful. Dealing with the subsidiary problems may not be as dramatic as dealing with the final, ultimate ones in dispute. Resolving them, however, will be a major contribution to the process of dispute resolution. It will clear the air for the final, public-policy decision.

"Scientific Consensus," Not "Negotiated Agreement"

The conference should seek only to find scientific consensus regarding scientific issues. It should not attempt to "negotiate" or otherwise create agreement on such subjects by accommodation.

Scientists have the right to claim expertise in answering scientific questions. They do not need to represent anyone to find consensus. They are not representatives or legislators, however. Where they seek to negotiate an agreement, they assume a representative or legislative role and necessarily must be including something other than strict science in their work.* The remainder of society has not assigned them such a function, and they have no special right to claim it. If they usurp it, they will impair the effectiveness of their scientific consensus-finding effort because they have overreached.

This avoidance of the legislative function, also should do much to encourage participation, and to enhance the credibility which attaches to the finding of scientific consensus.

Procedural Fairness in Appearance as Well as Fact

The conference administration must be as clean and pure as Caesar's wife. Not only must it be fair in everything it does, from selection of participants to assignment of hotel rooms and orders of presentation, everyone must perceive it to be totally impartial.

Both 2,4,5-T conferences were perceived as loaded in favor of continued 2,4,5-T use. There were comments that NCPP was biased in industry's favor, and many in industry believed that it was biased toward environmentalism. Some of this is probably unavoidable. Industry will usually have the greatest number of scientists working on the pertinent problems and, therefore, the greatest number of scientific participants. Industry also has the principal incentive to sponsor affairs of this kind and the resources to fund them. The result can be overweighted industry participation, which discourages attendance by others and thereby aggravates the matter. Environmentalists see all this as reflecting pro-industry bias. Industry, in contrast, sees the effort to bring balance into the picture, by special funding or other consideration accorded environmental representatives, as reflecting pro-environmental bias.

* A scientist reviewing an earlier draft of this manuscript, wrote: "The terminology 'strict science' implied an absolute fact when in fact, there are many instances where a 'scientific judgment' allows for broad generalizations (positive or negative) even in the absence of more specific detail. Most of the current problems involve such a massive amount of laboratory investigation that certain generalization may be possible based perhaps upon something less than, for example, a detailed investigation of every single member of some homologous series of compounds. With three or four benchmarks all showing a consistency one with each other, there is not the need for a detailed examination of all the others. Scientific judgment in such cases might well be a legitimate (valid) input to the scientific community." Private letter, October 5, 1979.

Wherever possible, the conference administration should include a skilled, nonscientist chairperson, such as Frank Quinn. Scientists are not different from other people. Even if they are seeking consensus and not negotiating, there will be inevitable initial reluctance, resistance, and posturing. An impartial chairperson can do much to break through all that, and to bring the participants to grips with their science. At least where there is such a chairperson, and preferably in all cases, it would seem wise to provide an independent secretariat and administration, such as CSIS furnished NCPP.

Broad Scientific Participation and Self-Selection

The conference must encourage the widest reasonable scientific participation, so that all shades of opinion are reflected. No responsible scientist should be excluded. There must be broad invitation and wide publicity. There must be ample lead time, so that other commitments do not interfere. Funds for travel, other expenses, and even lost time must be available. Existing scientific information should be circulated in advance, so that no one is required to undertake independent research if he does not wish to.

The finding of a scientific consensus requires that all "responsible" shades of opinion be examined. Broad participation by scientists in the conference is accordingly essential. The word *responsible* is not intended as a limitation, other than that the person be willing to participate and reason with other scientists as they reason with him. Scientists at the leading edge of current thinking may not be excluded—at one time or another, Galileo, Copernicus, Darwin, and many others were rejected by their scientific peers because they were too far in advance of their times. Yet ultimately their views became the consensus. We can not afford to repeat the same mistake.

Participation in the conference should be by "self-selection." This means that conferees decide for themselves whether to attend and do not need any permission or special invitation. It is in accordance with scientific method. It helps insure that subtle bias does not enter the selection process. It enhances the possibility of a conclusion being accepted as a true consensus.

Substantial lead time makes for better planning and preparation. Its absence discourages participation. The lack of adequate lead time was one EDF objection to the 1979 2,4,5-T conference:

> Robert Harris of the Environmental Defense Fund, one of the opposing scientists not at the conference, agrees with Tschirley that there was no organized boycott of the conference. He blames short notice for the lack of participation by opposing scientists. Most invited participants received notice of the conference four to six weeks before it was held.
>
> "You might as well organize the conference a day ahead," Harris says. "These people are busy."[37]

Funding of travel and other expenses, and possibly even of lost time, is similar. Environmental organizations assert that scientists aiding them do not often have independent financial support for conference-going, unlike their government and industry colleagues; this prevents their participation. Although participation in a consensus-finding conference should be recognized as a public service and not a money-making endeavor, it would be unfortunate if potential scientific contributions, worth far more than the cost to bring them to a meeting, were lost on financial grounds. Accordingly, funding of travel and living expense at the conference should be available as a matter of course to those seeking it, and for time lost from other income-producing activities upon some fair basis of a showing of need. These should all be held to a minimum (e.g., lowest cost travel), and the location and arrangements for the conference should be such as to make clear to all that it is not a junket.

The circulation of all pertinent scientific data and opinion well in advance of the conference will insure that everyone has the same data base, and again encourage participation by limiting the need for individual scientists to conduct research and investigation on their own if they do not choose to do so. After all, the very purpose of the meeting is to find the extent to which there is a consensus on the present state of knowledge, not to develop new information so that something better is created. Scientists who wish to should be able to participate simply by studying what is furnished to them, and by coming to the meeting.

If the time, effort, and financial obstacles are removed, a large scientific representation should be possible. Many scientists are most interested in—and involved in—public issues. They are accustomed to attending conferences. They are concerned citizens. If the fact and appearance of bias are eliminated, their environmental and other organizations will at least not boycott the conference or stand in the way of their participation. Indeed, if convinced that the effort to achieve true scientific consensus is sincere, one would hope that all segments of society involved in the dispute would welcome the chance to see whether there are real answers.

Wide Public Participation

A final guideline is that the conference must solicit the largest reasonable participation of all segments of the public and of the media.

The conclusions of the consensus-finding conference will not ordinarily be operative in themselves, but only insofar as they are factored into the public-policy dispute-resolution process by its arbiter. That may be a legislature, an administrative agency, a court, or some other dispute-resolution mechanism; ultimately, in the socio-scientific dispute, it will be the public at large. The value of the consensus, when discovered, is thus significant in direct proportion to the extent, and credible method, of its communication.

Credibility suffers in today's adversarial atmosphere of suspicion and hostility if there is secrecy and even the hint of "star-chamber" proceedings. "Sunshine" in matters affecting important public policy has become as much the norm in science as it has in government rule-making and administration. Moreover, lay observers can sometimes furnish valuable inputs even to scientists, because few scientific questions are truly free of all "value." Lay persons can recognize the policy issue sometimes, even if the scientists do not. At one National Academy of Sciences discussion, for example, one well-known scientist was stating his view that the levels of ortho-toluene sulfonamide (OTS), an impurity suspected of causing cancer, "should be as low as possible" in saccharin. It was only after a lay person pointed out that removing the last OTS molecule might cost $20 million, that he stated that what he really meant was to reduce OTS to the lowest reasonable level. This "would be on the basis, I think, of good manufacturing practice, that practice by which the lowest possible level of OTS can be achieved and still produce a commercially feasible product."[38] In turn, this, of course, indicated that the real issue was another risk/benefit economics one: What is "commercially feasible"? How much profit should the manufacturer make? What should be the consumer selling price?

Lay observation enhances lay understanding, which is important to resolution of the ultimate policy problems involved. Lay interest encourages media interest by suggesting that what is happening is "news" and should be covered. Media interest, of course, is independently significant by virtue of its public and political impact—and its influence on the conferees themselves. As NCPP showed, it heightens their consciousness of the public interest they are serving and encourages their efforts to arrive at results which serve the public interest. The combination of lay and media presence encourages some scientists interested in public affairs to attend. It also helps relieve what might otherwise be an anxiety or uneasy feeling among nonparticipants affected by the conference that something important could have been missed or suppressed. It thereby enhances the chances for wide public acceptance of the results. Lay, media, scientist, and public interest thus constitute a kind of round robin, which builds upon and feeds and improves itself.

Some scientists, of course, are reluctant to include outsiders in their scientific discussions. This is not only the result of the ivory-tower syndrome, but because scientific terms and concepts often have very different scientific meanings than the apparent lay equivalents. A phrase spoken quite properly during a scientific conference and then later repeated out of context on the front page of a local newspaper, or in an adversarial brief, can appear cold and callous, misguided, extremist, or even ridiculous. One day, when and if the credibility of the scientific consensus-finding mechanism has been established fully in the public opinion, it may be possible to limit some outside lay attendance. But even then, I fear, the reluctant scientists must resign themselves to a large measure of continued public presence. Their conference has no legal powers at all. Its sole effect is by virtue of the credibility and persuasive influence of its findings on adversaries, courts, administrators, executives, legislators, and, finally, the public generally. Communication to, and with, all these groups—which can number in the millions of persons—is as much a part of modern science as "science" itself.

The conference administration should pay special attention to media involvement. Journalists and writers should be encouraged to become involved intimately, so that they will understand what is happening and have a "stake" in communicating the information they have developed. "Public information" is an important specialty and discipline all its own, deserving of an important conference staff position.

The Local Conference

Public interest and concern understandably have focused on the major national and international socioscientific disputes, such as those with regard to nuclear power and suspect carcinogens. Most of these issues also have their state and local counterparts. Fluoridating the local drinking-water supply is a good example. With time and perhaps some reversion to more decentralized government, such local problems will burgeon. Ultimately, local socioscientific dispute resolution may become the most difficult problem of all. To the extent possible, any sound socioscientific dispute-resolution mechanism should be as applicable to local problems as it is to national ones.

The 2,4,5-T and coal-use conferences dealt with national problems, of course. But the consensus-finding proposal to which they gave birth should work just as effectively on the local level. It does not require any special governmental powers, or license or other grant, or participation by any particular national institution or organization of science. Citizens can administer it with scientists from their own communities.

The City of Denver's resolution of its dumdum-bullet problem was not exactly the kind of consensus-finding approach I have described, but it does illustrate how a similar effort worked on a local level. In 1974, the Denver Police Department decided to change its handgun ammunition to hollow-point (dumdum) bullets, because the conventional round-nosed bullet provided inadequate "stopping effectiveness." The American Civil Liberties Union (ACLU), minority groups, and others opposed the change, claiming that it would result in excessive injury to bystanders as well as targets. At first, the issue was fought publicly, with all the adversarialism that so often characterizes such debates. Then a number of scientists were called in and given the assignment of separating technical and scientific from social value judgments, and of integrating the nonvalue judgments analytically, not judgmentally. They achieved their objective so well that one of the bullets was accepted without further controversy by the city council and all the concerned parties. *How* they achieved it calls for considerable further discussion and is not relevant to the analysis here. The important point is that local scientists were called in to deal with the scientific aspects of a local problem. They did so to the satisfaction of all concerned.[39] The process would seem equally applicable to other local socioscientific issues.

Conference Funding

One final mundane, but critically important, issue remains for discussion—the matter of consensus finding conference funding. The value of the services to be furnished by the scientists participating in any such conference is enormous by any test. It far outweighs any costs which are incident to conduct of the meeting. Nevertheless, the holding of the conference does cost money, which most scientists cannot be expected to contribute. There must be a secretariat and administration to make arrangements, handle communications, distribute materials, undertake research, and generally administer the great many other activities which are incident to a meeting. Workshop and other session facilities must be hired. Travel, hotel, and living expenses, and some expenses for lost time, must be reimbursed. Reports of the conference, and "proceedings" of the discussions and rapporteurs' observations, must be published and distributed.

Governments, private foundations, and other groups do sometimes have funds available to support scientific conferences. One day, when and if the new approach has been established and perhaps institutionalized, moneys may come from these sources, or even from an aggressive and well-organized citizen effort, or from some combination of the two. For the present, however, the really important backing sufficient to test the process adequately will probably have to come from industry, which carries a heavy burden of proof in most socioscientific disputes.

As the first 2,4,5-T conference showed so dramatically, where there is a litigation pending there are clear and very significant trial-preparation benefits to a scientific conference which concentrates exclusively on the issues in a socioscientific dispute. Even though it seeks scientific consensus as the fundamental objective, it also gives participating scientists the equivalent of a dry run—an opportunity to test their research and opinion with colleagues in order to see whether there are defects which might have escaped notice. It encourages scientists to participate in the dispute-resolution process when otherwise they would be reluctant to leave their scientific ivory towers. It demonstrates an openness which is persuasive court evidence of confidence in the merits of the contention. It encourages and helps marshal third-party support, which might be difficult to find. Through the pressure of peers meeting with and questioning each other, it attenuates opposition extremism, and develops an atmosphere in which accommodation is more likely. Where attorney adversaries stand between adversary principals, it furnishes a direct pipeline to the top. Where political pressures are great, it may provide a government executive or administrator with the excuse he needs to adopt an otherwise unpopular course. Where it finds consensus, it eliminates areas of dispute. It uncovers defects in a position before the position is frozen, and in time for correction. For these and many other reasons, it is a tremendously useful trial-preparation tool. Indeed, the legal and witness expense saved by "preparing" scientists together, at one time and in one session, can alone be many times the cost of the conference. Industry support of a consensus-finding conference in the relevant area, and at the right time, would thus seem justified as a matter of legal budgeting alone.

Even where litigation is not pending, there is still reason in most cases to hope that industry will support the consensus-finding approach. Except for the truly political issues, such as inflation and SALT, it is almost always industry's products and services which are being questioned, and industry's science and scientists which are under attack. Industry is losing a far greater proportion of the battles than its science would seem to justify. However large the costs of a scientific conference may be, they are quite literally trivial by comparison with the real stakes involved. One would hope that any business committed to long-run profit maximization would not only be willing to make the necessary small investment, but would, in fact, welcome the opportunity if it is presented.

It must perhaps be added, out of caution, that industry funding must be on a totally "no strings" basis. Any effort to limit payments to specific persons or in connection with particular issues—outside of the broad consensus-finding charter of the conference—could easily lead to an adversarial perception by some that there has been an attempt to bias or influence.* Then the expenditure will not only have been wasted but the integrity of the claim will be damaged, and the effort to test the scientific consensus-finding process rendered worthless.

There remains an important group of socioscientific disputes which cut across society generally. It would be good, of course, to have credible scientific guidance with regard to the technical issues involved in the trade-off between inflation and unemployment, or the possibility of insuring Russian compliance with SALT. But we must also crawl before we walk. It may be best that there is no apparent source of support for such an effort at this early time, for experience in dealing with such broader societal concerns is even more lacking than elsewhere. Indeed, it might well be the wiser course to begin the consensus-finding test process on the strictly local level, where mistakes can be corrected with minimal harm either to people, or to the process itself.

* The view of many persons is probably well-summarized by the scribbled marginal notation of one scientist who returned an earlier draft of these pages, "He who pays the price calls the tune." I am not so cynical, but certainly do agree that a major effort is required to avoid such a result.

8 A PLEA FOR UNDERSTANDING

On July 31, 1979, four months after the Three Mile Island episode, Dresser Industries placed a two-page, center-spread advertisement in the *Wall Street Journal*. It featured a quarter-page photograph of Dr. Edward Teller, the nuclear physicist who has led much of the struggle for commercial use of nuclear power. A bold-type, double-page headline reading, "I was the only victim of Three-Mile Island," continued with the lead paragraph:

> On May 7, a few weeks after the accident at Three-Mile Island, I was in Washington. I was there to refute some of the propaganda that Ralph Nader, Jane Fonda and their kind are spewing to the news media in their attempt to frighten people away from nuclear power. I am 71 years old, and I was working 20 hours a day. The strain was too much. The next day, I suffered a heart attack. You might say that I was the only one whose health was affected by that reactor at Harrisburg. No, that would be wrong. It was not the reactor. It was Jane Fonda. Reactors are not dangerous.

The balance of the two pages argued the case in favor of nuclear power.

Two weeks later on August 16, the *Wall Street Journal* carried a letter in response from NRC Commissioner Peter A. Bradford, whom some in industry consider to be an "environmental" adversary. He implied that the ad was misleading, because it failed to disclose "that it was a Dresser valve that stuck open during the Three Mile Island accident sequence, allowing the cooling water to escape the reactor cooling system and exposing the nuclear core." The *New York Times* picked up Bradford's charge the next day in an editorial entitled "Propaganda." It ended:

> There is another noteworthy omission in the Dresser ad. A valve stuck open at Three Mile Island and contributed to the accident. It is not yet known whether the malfunction was in the valve or in an improper signal relayed to it or whether something else went wrong. But it is at least possible that the manufacturer of the valve will turn out to have been partially responsible for the near-disaster. It would have been nice of Dr. Teller or the advertiser to have identified the manufacturer of that valve: Dresser Industrics.

The exchange continued the following week, with the *Journal* publishing a letter on August 24 from James R. Brown, Jr., executive vice-president of Dresser Industries. It denied that the valve was the cause of the problem. It contained an attack on Commissioner Bradford, as follows:

> What Mr. Bradford did not reveal to you or your readers is that he is a former member of Ralph Nader's staff. Also, he did not mention his own bias against the very industry he is supposed to regulate. As the National Journal pointed out, of the three persons named to the Commission at that time, "Bradford is considered to be the most skeptical of the nuclear industry."

This is where adversarialism has brought us. Whatever light this public debate might have cast on this very complex scientific dispute was quickly extinguished by *ad hominem* charges of improper motivation, willful suppression and obstruction, and deliberate misrepresentation. The public does not know whom to believe. It is little wonder that so many people are disgusted.

The first two chapters of this book have sought to analyze socioscientific dispute resolution to set the stage for the development of new and better approaches. The following chapters have presented a number of specific suggestions and have encouraged others to consider new measures of their own. In contrast to dispute resolution by accommodation, these new approaches are not necessarily predicated upon agreement between adversaries. Nevertheless, proposals would be far easier to develop, test, and implement if the contestants at least trusted each other and acknowledged that their opponents were persons of good faith who sincerely want to achieve what they regard to be best for society.

I am convinced that substantially all of the present adversaries to socioscientific disputes are genuinely searching for what they believe to be in the true public interest. They may be adamant. They may be obstinate. They may be wrong. But they are driven by a real sense of public purpose. If their opponents can only be made to understand this, much of the heat and bitterness can be eliminated from the contests. Then both sides can move more responsibly and rationally to serve the public.

This final chapter seeks to encourage that understanding between adversaries, so that the long trip toward improved societal control of science and technology may move more rapidly forward. It is addressed largely to industry, which is losing so badly. Industry *must* change its view of the opposition, as a simple matter of survival. Its present course leads to sure loss. But the chapter is addressed to both sides as well, in the belief that most of the bitter adversaries are decent people and organizations, struggling to improve life on this planet. We will all benefit if they can be made to see each other that way.

"Industry" vs. "Environmentalism"

The kinds of socioscientific problems that society today faces make it virtually inevitable that industry and its environmental and consumer opponents will oppose each other vigorously on many issues. But the present struggle has much of the religious fervor of a holy crusade. Large numbers of people on each side believe that their adversaries are inherently bad, and are seeking to destroy society, or are motivated exclusively by avarice. Each side believes that the other will cheat, lie, steal, and (as in the implied Silkwood[1] charges) even kill to win. Certainly some of this may be true. One cannot involve millions of people in important disputes with huge stakes without anticipating some misconduct and even crime. But to predicate a broad, societal movement on such presumed motivation is wrong, and in the end must necessarily disserve that society the movement seeks to serve. I think it can be demonstrated that such presumptions are unfounded.

Labels are usually dangerous, but sometimes (as here) there is just no way of avoiding them. The key present socioscientific disputants are "industry" and "environmentalism." "Industry" signifies those who shade the risk/benefit balance in favor of continued or expanded utilization of technology. Their purpose is to produce the material goods and services which they believe will in turn relieve poverty, improve health, and result in a better society. "Environmentalism," which includes "consumerism," means those who shade the balance the other way. Environmentalists believe that dangerous side effects of technology threaten to destroy emotional and esthetic as well as material values. They believe that uses of technology which pose risks are unacceptable unless those favoring such uses sustain a burden of showing that the benefits outweigh the dangers. They shade the balance against continued or expanded utilization of technology. The differences between the two groups are thus very largely with regard to burdens of persuasion.

Most readers will be familiar with the general run of attitudes, charges, and countercharges of each of these groups; many readers probably identify with, or are members of, one of them. But few persons who are not directly and intimately involved in the struggle appreciate the depth of emotion and feeling shared by large numbers of decent, responsible, and trustworthy advocates on both sides. Adversarial responses to the "rule-of-reason" proposal are most revealing in this respect.

The rule of reason is hardly an extremist weapon. The worst one would expect its critics to say is that it is "utopian," "unrealistic," "a waste of time and diversion," or "naive"—not to charge that it is dangerous and immoral as well. By quoting such far stronger reactions below, I do not mean to coat all industry or environmental advocates with the same tar. Of course there are gradations of opinion, with different people having different views on different disputes. But the underlying emotional response is so often the same as to smack of the Arab-Israeli conflict, and to make one wonder, "How can grown people act like that?"

A successful trial lawyer, who has published, taught, held public office, and generally earned the esteem of his profession, categorically rejected the rule-of-reason approach with the following diatribe:

> Corporate rapacity has not diminished one whit since Professor Adolph Berle taught us in law school in 1946 that the new class of corporate managers might someday act as enlightened fiduciaries.
>
> Multinational oil companies have taken advantage of the Arab oil embargo to maximize their profits in partnership with Arab feudal sheiks at the expense of the United States.
>
> Other blue chip outfits were up to their eyeballs making illegal contributions to Nixon's campaign.
>
> And how about their pervasive practice of bribing foreign officials in exchange for lucrative war contracts (the modern merchants of death).
>
> Not a day goes by that one doesn't marvel at the vast spectrum of corporate crime and subversion, domestic and foreign, ranging from CIA dirty tricks to selling contaminated ice cream.
>
> When it's ethics vs. the buck—you know as well as I do which way big corporate enterprise will bounce.

A prominent attorney from the opposition camp read a description of this attorney's adversarial tactics "with horror," characterized his conduct as "nefarious," and ended up calling him a "skunk"![2]

An industry advocate wrote to *Chemical Week*, protesting its editorial endorsement of the rule-of-reason approach as employed in the National Coal Policy Project:

> If you . . . feel anything will or should come from this NCPP idea, you are incredibly naive. As a . . . generally vocal antienvironmentalist, I have dealt with eco-freaks directly and indirectly, and can tell you the following about them:
>
> —They believe God is on their side. (Ever try to discuss religion with someone?)
>
> —They do not acknowledge your right to exist.
>
> —They clearly recognize you as the enemy, and act accordingly. Unfortunately, we don't.
>
> —While we talk about controlled growth and sharing available resources, they are thinking about how to eliminate you. . . .
>
> "Rule of reason" has no meaning to an eco-freak. What can you do? As a starter, don't write editorials that imply that eco-freaks are reasonable (they are not). Instead, exhort the troops to go out and fight them. If you want a good example of how to do it, pick up a copy of the "Sierra Club Bulletin" and just reverse everything. One word that never appears in it is, *Compromise.*[3]

Chemical Week is a widely circulated chemical industry trade journal, which will publish only material which it considers worth communicating to its industry. Later it published another letter from this same writer, who spoke of "our never-ending war with public enemy number one, the environmentalist . . . who hates people and wants everyone to freeze to death in the dark."[4]

Leaders of each side often have an almost messianic conviction of the merits of their own positions. They detest and even hate their adversaries. The result is that both sides frequently resort to "the end justifies the means" tactics described by Kuntsler in chapter 3. Then each, seeing what the other does, becomes even more persuaded of the lack of integrity and dishonesty of the adversary. The process becomes an almost self-perpetuating one.

Many in the environmental movement consider that industry is almost literally stealing stockholder assets when it employs corporate funds for advertisements, general publicity campaigns, lobbying purposes, and other such activity in support of an industry position in a socioscientific dispute. They fight back with tactics designed to maximize their own far-more-limited resources by stirring up the media and thus overcoming the inertia of the traditional dispute-resolution mechanisms. Their moral justification, of course, is industry immorality, the grave danger to society (which can justify even revolution), too little time for ordinary processes to work, inadequate funds and other resources, and the need to do something. Their tactics include hyperbole, premature and inadequately considered public charges, vilification, and marginal and provocative legal "tricks."[5] Their dispute-resolution results are better than industry's, but the damning consequences to scientific credibility and sensible dispute resolution are disastrous.

The Industry Response

Many in industry believe that their own serious losses, and the opposition wins, are a direct result of these "irresponsible" public charges, environmental media "manipulation," and "questionable" legal tactics. Their natural response, also morally self-justified by the perceived immoral conduct of the adversary, is legal trick and "PR" of their own. The adverse consequences to them are even more disastrous than those resulting from environmentalism's tactics.

I have already referred to the potential for industry's adversarial extremes backfiring and impairing industry's contentions. Many in the environmental movement understand this and quite deliberately employ adversarial extremes in order to provoke just such industry tactical extremes. In discussing litigation tactics and strategy, *Rule of Reason* described this process as follows:

> Indeed, it is not uncommon for the "consumer" opposition to itself employ the sporting approach, hoping to provoke an industry adversary into ever more open and flagrant use of the same tactic. The objective is then to lower the boom and charge that such conduct demonstrates industry's fear of the facts. It is a taunting approach and extremely effective, especially among those to whom "turnabout is fair play" has become dogma in litigation. Turnabout may be fair enough play, but it can also produce disaster. The sporting approach is not only industry's soft and weak underbelly; it is being used by the opposition as an affirmative weapon to pierce that underbelly and destroy the beast (p. 19).

Industry's track record in "fighting fire with fire" by employing adversarial legal extremes of its own has been bad enough. Its effort to match environmentalism's public relations achievements has been far worse. Too many in industry have interpreted pleas by industry leaders for them to take industry's case to the broad public as calls for industry "PR," in the most unfortunate perjorative sense of that term. The result has been serious damage to industry credibility.

Anthony Harrigan, speaking as executive vice-president of the United States Industry Council, a nationwide association of conservative businessmen, put it this strongly: "If economic freedom is to survive in this country, businessmen must fight back in the forum of public opinion."[6] The response to calls of this kind has been substantial. Early in 1979 (January 22), *Business Week* published one of its lengthy "Special Reports," entitled, "The Corporate Image," and subtitled, "PR to The Rescue—For the company under siege a bigger role for public relations." More and more large companies have introduced internal public-relations education and training programs, designed to teach executives how to deal with the media, and with the public. In a mid-1979 Sunday edition (August 26), the *New York Times* commented on this effort in an article entitled, "Executives Taught to Improve Images at Seminars."

The educational purpose may be laudatory. But the way in which it has been implemented has too often been the opposite. Irving Kristol, Henry Luce Professor of Urban Values at New York University, and one of the best known commentators on corporate ethics and responsibility, protested:

> Essentially, as I see it, the problem is one of candor and credibility, not—repeat: *not*—of "public relations." Indeed, one of the reasons the large corporations find it so difficult to persuade the public of anything is that the public always suspects them of engaging in clever public relations, instead of simply telling the truth. And the reason the public is so suspicious is because our large corporations so habitually do engage in clever public relations instead of simply telling the truth.[7]

John W. Hill, a leading public relations executive, argued even more forcefully:

> An adversary situation has developed between corporations and the media. Business thinks the media are hostile toward it while the media think business is unresponsive and secretive. . . .
>
> Desperately needed is a whole new thrust toward corporate candor. Top management can no longer shirk its responsibilities for synchronizing and linking the vital elements of policies, performance and communications. Deceptions, nondisclosures and obscure double talk must give way to openness, forthrightness and clarity in matters of public concern.
>
> If public opinion is misinformed, misled or lied to, it can be a destructive force. Business must show, by policies and acts in the public interest and by speaking out clearly and convincingly to people, that it is worthy of their support and confidence. *In my opinion, the survival of private enterprise will depend on how well this job is done.* (italics added)[8]

Industry cannot adopt the tactics of its adversaries, whether those are proper, as environmentalists consider them, or immoral, as industry perceives them. The public simply does not consider industry to be an "underdog," or accord it the justification of necessity by virtue of limited resources. Industry has no real choice but to "turn the other cheek" to adversarial extremism in any form—however painful it may be. It must employ the rule of reason in all its conduct, and act in what it regards to be the true public interest, rather than striving simply to defeat or destroy its adversary.

The pain of rejecting weapons employed by an adversary is undoubtedly exacerbated by the conviction that the adversary is bad and therefore must be defeated. If the adversary can be seen as only trying to win a cause, and not as bad in any moral or ethical sense, some progress in bringing logic to the socioscientific dispute-resolution process can be made.

The "Intent" Fallacy

Why is there such powerful conviction in environmental vs. industry conflicts that the adversary is bad and therefore to be hated? I believe that one important reason is because "intent," as we have discussed earlier, is so frequently an important legal issue in these matters. The parties assume that "bad intent" means "bad person." But this is a fallacy. The "intent" which is involved in these disputes is a legal fiction that has nothing at all to do with what goes on inside the other side's head and heart. If adversaries can only be made to understand this, perhaps some of the venom and animosity can be removed from these disputes.

The intent involved in the socioscientific dispute is "legal intent," not "moral," or "ethical" or even "inner man" intent. A "bad" legal intent has no necessary connection at all with morality or ethics. "Legal intent" is the motivation discussed in chapter 1. Where an organization is involved, it is always a fiction. Organizations simply do not "think" like humans. Even where human beings themselves are concerned, legal intent is very often equally a fiction. It can be the precise opposite of what the individual "believed," or "experienced," or "felt." Whatever goes on inside a person's head—provided he is not insane or externally coerced, of course—an individual is legally responsible for what he does. If he makes a conscious choice between something which is lawful and something which is not, he has the necessary "legal intent" for purposes of assessing responsibility. It matters not a whit, for this purpose, that he was motivated by what some might consider to be the most commendable purposes, such as Robin Hood's objective to serve the poor, or John Wilkes Booth's purpose to end tyranny. An eminent psychiatrist, Walter

Reich, explained it well in discussing the problem in connection with the Patricia Hearst case:

> Criminal law is based on the assumption of personal responsibility for one's behavior, and this is in turn based on the assumption of free will. Human beings can choose between right and wrong; if they do wrong, they can be assumed to have chosen to do so, and if they have done so, they must be held criminally responsible.[9]

Legal intent is thus not a proper factor upon which to predicate animosity. If the emotional response of hatred is ever justified—and I believe that it is not, although I too sometimes have feelings—it is "inner" intent which counts. One should not hate because of what another person does, but because of what that person is, or thinks, or feels. Even here, however, I think most people would agree that the better medical view today is that there is no such thing as a "bad" inner intent. The "Son of Sam" killer thought he heard voices, telling him to kill. "Hitler's heart," as one psychiatrist suggests, "was in the right place."[10]

Father Flanagan, the humanitarian priest who created Boys Town as a refuge for wayward children, had a favorite saying, "I never met a bad boy." Psychologist Carl Rogers, whose teachings are widely acclaimed, believes that "people are basically good."[11] In the "inner man" sense in which both were clearly speaking, I am convinced that they are right. I have been a trial lawyer for three decades. I have been a prosecutor and defender of criminals; an advocate and opponent of industry and environmental forces; a representative of husband against wife and wife against husband. Some of my cases have been long ones, in which I have literally lived with a client. People have bared their inner souls to me, sometimes far more than to their own closest relatives and friends, or even to themselves. I have met many people, including clients, who did "bad" things. I can honestly say, however, that I have never met a "bad" person, once I have really gotten to know him.*

* This almost inevitable "understanding" of the client in a lengthy litigation can have its negative implications as well. Many years ago, I was representing a defendant in a major criminal case. My then senior partner, James S. Hays, took me aside to point out that my increasingly close and sympathetic relationship with the client was threatening my ability to deal with the facts independently. This in turn could jeopardize communication with the judge and jury. The lesson was an important one.

I recall particularly one eight-week trial in 1959. I was serving as special federal prosecutor of the twenty-seven so-called "Mafia delegates" to the Apalachin, New York "crime convention." It could not be proved, but there was good reason to believe that one of the defendants was the national leader and "godfather" of the group, who had engaged in a lifetime of crime, including brutal "enforcement" murders. He was a small, elderly gentleman with a kind face. Sometimes he would stop me in court and draw from his wallet pictures of his wife and children and grandchildren, and of his meetings with respected religious leaders. Real tears were in his eyes. He was not trying to influence me—he did the same to others, including court "buffs," and anyone who would listen. He was a deeply religious person and a good father and friend. He was hurt by this public experience. He knew that whatever he had done was because it was "right" for the society in which he had been brought up and lived; he could not accept the rules of a very different society, which he knew to be oppressive and unfair. I was convinced of his guilt on the evidence as I knew it; I also believed his inner motivation to be what he "knew." I could not hate him.[12]

"Legal intent"—derived from what people and organizations do—is all that industry and environmental opponents should examine. If they do so, they need never hate each other—"sticks and stones can break my bones, but words will never hurt me." If adversaries must consider inner motivation, they will find that it is good and, once again, need never hate. And there is one final reason why both sides must abjure hatred. It is a destructive force, which consumes and tears apart self even more than its object. It interferes with rational pursuit of purpose and goal. Most important, it disserves the public by focusing attention upon an erroneous irrelevance, and substituting shout and heat for logic and light.

What we do is what we are, not what is inside us. It is all that really counts. Let's fight about that alone.

Our Societal Crusade

We *are* in a crusade. But it should not be a holy war between industry and environmentalism. It should be a struggle between risk and benefit. It should be an effort to understand just how we can best employ the wonderful and exciting new things science offers us, in the interests of all. It is fitting and proper that the adversaries in the struggle battle for what they each believe to be best. Their fight will help to enlighten all of us, and thereby lead to the optimum decision. But they must fight in a manner to achieve that enlightenment, not to further obscure or mislead.

Opponents must not resort to adversarial extremes, no matter how much they consider them justified by opposition tactics. Nor must they yield to what they perceive to be such tactics, and give up in despair, or "cop out" in disgust. Both sides may be losing some struggles, but both sides are winning others. Industry does have a poor batting average in socioscientific dispute resolution, but it has heard a Democratic "populist" president say that "inflation is our most serious problem." It has seen the beginnings of real action to achieve greater productivity, budget balancing, lower business taxes, and deregulation. Environmentalism has of course seen the same general governmental trends, but it has also become a powerful force in our society. It has already won for itself as effective a place in the "establishment," as the labor movement gained only after a much longer and far more difficult struggle.

The battle must be over reality, not figment. It must be based on data and logic, not calumny and invective. Opponents must give each other the benefit of good faith and proper motivation. Then, however misguided some conduct may turn out to be, each will be able to extract what is constructive and good from the other.

Despite the vigor of so much of the battle, things are comparatively quiet in American society today. Gone for the moment are the Vietnam and Watts riots, the civil rights marches—and even some of the cries for equality. But do not let the quiet mislead. It is like the calm at the center of the hurricane, the lull before the storm. We have time to do what is needed, I believe, but we do not have forever. Dissatisfaction, disaffection, alienation, and worse are still there beneath the surface, waiting to erupt. Few persons are satisfied with our mechanisms for resolution of the major differences among us about "quality-of-life" issues—how we live, where we are going, discrimination, equality, the environment, poverty, peace. There is cynicism, doubt, and downright disbelief in our political and business leaders, our professionals, our courts, administrative agencies, and legislatures—and just about anyone and anything involved in the leadership of society. For once, this dissatisfaction is expressed as much by members of the "establishment" as by the outcasts and pariahs. Indeed, corporate society is sometimes even more outspoken in its condemnation of government than are the "public-interest" spokespeople.

Our leaders speak in grand terms of "human rights" and the "war on poverty." Few expect the answers to come quickly or easily, but we do have a right to expect that our leaders' performance in pursuing these stated grand objectives will be consistent with the objectives themselves. They must try to do what they say they will do, or what they say will be rejected. Unfortunately, however, all too frequently, conduct and stated objective appear inconsistent. The official protesting the need for forthrightness seems to hedge—and we no longer believe him. The executive promising a quick decision appears to delay—and we no longer believe him. The legislator calling for integrity condones disingenuity—and we do not believe him. There may be many causes for the present alienation of so many persons, but this inconsistency between promise and conduct is a critical central one. The two must track. They must match. The rule of reason demands that they do. Its use by our leadership, and by all those involved in resolving the important issues, will go a long way toward putting us on the right problem-solving track.

The "futurists" tell us what our new world will look and be like a generation hence.[13] The range of their predictions is enormous, and no one can be certain of who is right and who is wrong.* But of one

* It is not strictly relevant to the theme of this book, and I claim no special expertise as a "futurist," but I cannot resist the temptation to add my general prediction of the future as one final note.

The enormity of the societal problems we face might lead to pessimism. But history makes me optimistic. Thomas S. Kuhn's *The Structure of Scientific Revolutions* (Chicago: University of Chicago Press, 1970), a remarkable description of the major scientific developments in our past, is persuasive evidence that there must be new discoveries ahead of us which are at least as dramatic as those of Newton, Darwin, Einstein, and the many others who have moved the world into new channels of thought and life. Whether it is the unified field theory (or theories), reproduction by "cloning," telepathic communication, extraterrestrial life, or something else, matters not nearly so much as that *something* new is ahead which will provide solutions to problems which today seem insoluble—and undoubtedly at the same time change society in ways which we cannot now anticipate. To conclude otherwise, and to believe that we have made all the really major discoveries there are to be made, requires an arrogance and conceit which there is little in human history to justify.

Our problems *are* great, and they *are* serious. The next discoveries may well pose even more serious and difficult new ones. But all this makes for challenge, and challenge makes life exciting and fun. Without worlds to conquer, this would be a dull life.

And while we're waiting for the next scientific genius to lead, it would be good for the rest of us to try to give society a little shove in the right direction.

thing we can surely be certain: whatever it is, the world will then be very different from the way it is today.

We can not let that change simply happen. We must manage the resources we have been given. We must shape the best possible world we can.

APPENDIX A

PUBLIC FORUM

COAL AS AN ENERGY RESOURCE: CONFLICT AND CONSENSUS

NATIONAL ACADEMY OF SCIENCES
WASHINGTON, D.C., 1977, pp. 258–60

Report of Dispute-Resolution Workshop, April 5, 1977

The national energy controversy, which is the subject of this National Academy of Sciences Forum, is one of a class of modern disputes characterized by several similar factors. First, and most important, it is a public interest dispute. That is, the absentee and largely uninformed public has a vital stake in the outcome of the issues being debated by the disputing parties. Second, it is a socio-scientific dispute. That is, the critical concerns are social, determining how we will live on this planet, and depend in major part on scientific judgments and analyses. Third, it involves extremely complex fact and opinion considerations in several highly technical disciplines, much of which is far beyond the competence of the lay person and aspects of which are beyond even the scientific expert skilled in a particular specialty. Fourth, there is no absolute answer, no right or wrong, no good or bad. Any decision must be based upon a balancing of a large number of pluses and minuses and of risks and benefits. Different people with different values will come inevitably to different decisions.

In a democratic society, in which ultimately the public must make the final decision, the combination of these factors means that credibility of the dispute resolution process is essential. Our dispute resolution workshop was agreed—as we all certainly must be—that our society has not yet developed such a process, which the public accepts as achieving optimum resolution of these issues. As one of the speakers at yesterday's plenary session said, "There is a crisis in dispute resolving." Cornelius B. Kennedy, a discussant at last night's workshop and chairman of the American Bar Association's Administrative Law Section, emphasized the historical inadequacy of legal and administrative processes, taking Dean Roscoe Pound's 1906 speech in this regard as his text. In present-day socio-scientific disputes, these concerns are intensified: Existing mechanisms are far too long and costly; large segments of the population are dissatisfied with the outcome of any case that they don't really understand and that, rightly or wrongly, they believe is determined by wealth, entrenched position, or activism, rather than merit. There is too much confrontation, with mistrust and hatred turning us against each other and putting too many people at the throats of others.

Not surprisingly, also, the workshop was unanimous that society needs to explore new approaches to resolution. It must find among those who are concerned with the present process, the leadership that will develop the necessary trust and confidence.

A number of such new approaches were explored. Dr. Amos Jordan, executive director of resource programs at the Georgetown University Center for Strategic and International Studies, described the National Coal Policy Project, which is bringing together opposing industrial, environmental, and other interests in an effort to reach accommodation—a sort of "environmental conciliation" procedure. The workshop was agreed that the earlier Forum discussion, rejecting mediation, was premature.

Richard Fleming, chairperson of the newly organized Chemical Industry Institute of Toxicology, described that institute's research effort to develop and publish scientific data in which the public can have confidence and on which it can rely. James N. O'Connor, assistant general counsel, Dow Chemical, United States, described two socio-scientific public interest cases in which the "sporting" approach of so much modern American litigation has been replaced by a "rule of reason," with resulting accommodations acceptable to all parties, and in the public interest.

In the discussion that followed, it further appeared that any new approaches to dispute resolution require the following as a minimum:

First, vehicles must be created so that accommodation can be sought among the presently enormous number of scattered and splintered views. If we are to achieve nonconfrontation resolutions, we must first have a workable number of units and spokespeople who can discuss, understand, and compromise. This was the thrust of the experience related by Dr. Jordan.

Second, the so-called "establishment" must accept the primary leadership role in reducing confrontation and introducing reason and logic into the resolution process. Presently, there is extremism at many points on the opinion spectrum. Whether or not extremism is ever justified—and remember, it was just last year that England as well as the United States celebrated the bicentennial of our own revolution—it may be at least understandable on the part of the deprived and disenfranchised when they have no other means of getting attention. The thrust of the experiences related by Messrs. Fleming and O'Connor is that the leaders of science, industry, and society must respond to the extremism of "outsiders" with data, reason, and understanding. There can be no excuse for the all-too-common response of equal or even greater extremism—such as the characterization in this Forum of one participant's environmental opponent's successful litigation conduct as "undemocratic," a characterization not dissimilar to the charge of "communist" a quarter-century ago; the refusal of one of the utilities, at least peripherally involved in the Kaiparowits case, to respond to the Sierra Club's several written requests for information, justified by some of the participants here as not the utility's responsibility and giving rise to the comment of an environmental leader, "We are sustained by the arrogance of our adversaries"; or the act of violence, however minor, which occurred on the floor of this great assembly hall at last month's Forum on Recombinant DNA in bitter reaction to the vigorous and outspoken but nonviolent and orderly demonstration of those opposing further research. All of this has contributed to the mistrust of industry, and science, and the establishment by a major segment of our society to the point where a good part of the public simply will not believe the unanimous assertions of the oil companies and industry regarding even this energy problem.

Third, as was pointed out by Robert White, director of the Academy Forum, the scientific community must intensify its efforts to enable the public to understand, or at least to evaluate and judge, scientific information so that public consensus or accommodation is possible. Reference in this connection was made to the usefulness of standards, professional accountability, certification of procedures, codes of ethics, and other specific measures in the effort to enable the lay public to deal with esoteric technical matters. The concerned lay person who traveled here all the way from Montana because of concern about the quality of life in her home community must be as satisfied as the most technically qualified energy expert. The public must be treated with dignity and respect—not patronizingly or with public relations gimmickry.

And, finally, the workshop was agreed that there must be a major effort by all to understand the views of any opposition and to accommodate to it wherever possible. The "scientific method" of consensus may be no more possible in many of these areas than it is between Russia and the United States, but accommodation *is* possible and may spell survival: People on one side or the other of a dispute are not usually "bad"—they simply have different perceptions and values, resulting from different environments and backgrounds. It is the unique spouse who can get through the first weeks, months, and years of living together without such understanding and accommodation. Society is no different.

Not long ago, our cities were wracked by riots, caused by the disaffection of major segments of our society. The socio-scientific issues discussed in these Forums have not yet generated the hatred or despair of the civil rights or Vietnam debates; but, unchecked, one day they might. It is up to those who pretend to leadership to see that they do not and that our nation continues in democratic fashion to search for and to find solutions to even the most troublesome and difficult problems, without confrontation, force, or violence, in the interest of all.

Milton R. Wessel, *Chairperson*

APPENDIX B

LEGAL FUNCTION MANAGEMENT PROPOSAL

Contents

I

Major Recommendations

The legal function of the Company should be organized and operated on the following basis:

1. *Internalization.* Only persons who are regular, full-time employees of the Company should handle recurring legal matters. Where outside assistance is required, as for a specialized *and* nonrecurring problem, Legal Department supervision should be close and continuing.

2. *Centralization.* All legal matters should be handled through the Legal Department. Outside counsel should be appointed by and be responsible to the General Counsel.

3. *Preventive law.* Anticipation and avoidance of legal problems should be a major Legal Department responsibility. The following counseling, educational and prophylactic programs should be regular Legal Department functions:

a. Claims, disputes, litigations, and unusual or abnormal legal experience should be administered by a single central review process, which serves as early warning to general management of potential problem areas.
b. Routine procedures should be followed to maximize trade secret protection, and attorney-client and work-product privilege.
c. Document control and retention policies should be administered as a part of regular record-review practices, designed to insure that the corporate documentary record accurately reflects corporate experience.
d. Written legal guidelines in major areas of legal concern, such as antitrust, environmental and civil rights, should be distributed to all appropriate personnel. These should be supplemented by regular oral presentations, or by videotape and videocassette showings where time or distance pose problems.
e. Corporate interdisciplinary teams should be appointed to manage important issues requiring the judgments of several

disciplines in addition to legal, such as compliance with §8(e) of the Toxic Substances Control Act (TSCA).

4. *Administration.* The Legal Department should operate on a budget, and with the same cost constraints as any other major unit of the Company.

5. *Personnel.* The minimum staff of the Legal Department should be the following:

a. Legal. General Counsel
Associate General Counsel (Commercial)
Assistant General Counsel—Patent
Assistant General Counsel—Litigation

b. Nonlegal. Legal Administrator
Claims Investigator

6. *Timetable.* The revised, expanded, and reorganized legal function should be introduced on a gradual basis, with current legal operations continuing generally in parallel until there has been sufficient experience with the new to be assured of quality and stability.

It should be assumed that no more than one additional new senior attorney will be employed each year, and that the new structure will accordingly not be fully implemented and operational until at least the fourth calendar year following this report, and possibly not until the fifth or sixth.

7. *Costs.* The new structure should result in a very substantial reduction in legal costs over the long run, were the same quality and level of legal activity to be assumed.

Over the short run, introduction of the new structure, indoctrination of new personnel, and the need for parallel operations, will undoubtedly result in some increased costs. These probably will be somewhat less than the first-year incremental cost of each new attorney, including associated overhead.

The same quality and level of activity cannot be assumed, however. The primary reason for these recommendations is improvement of quality, not cost control. If they are valid, the Company should have far better legal representation than otherwise, with resultant improvement in operations generally. Quite possibly satisfaction will result in an increase in the level of Legal Department activity, not only as a consequence of Legal Department initiative, but because of the wishes of nonlegal management.

For present purposes, and until better information and experience is available, a normal Legal Department budget of approximately $500,000 annually should be used for planning. This does not include settlements paid to third parties, or nonrecurring items such as seem to have doubled legal expense over normal during the first half of the current year. It does include all internal legal costs, and fees to outside lawyers, including those paid indirectly through insurance carriers, insurance adjustors, or collection representatives.

A number of other subsidiary or detail recommendations are set forth in the balance of this report, and identified as such.

II

Assignment and Scope

These recommendations are submitted in response to the Company's request for a long-range program for the operation of its legal function. They are to consider not only the Company's present experience and the regulatory environment, but also the Company's and its industry's long-term growth, anticipated changes in the government climate, and other relevant societal trends such as consumerism, environmentalism and populism. All of these have their inevitable impact on commercial operations in areas such as products liability or "corporate responsibility."

What is presented is accordingly a long-term plan. The Company's legal operations have been reviewed, but no effort has been made to analyze present activities in detail, such as by reviewing litigation files or conferring with outside counsel. Present costs and experience have been considered, but only on an "order of magnitude" basis. There has not been any significant attempt to ascertain the precise number of present and past workmen's compensation claims, for example, or the experience of each of the outside counsel who have handled products liability claims against the Company; responses to specifically posed questions, such as the scope of TSCA §8(e) compliance, have been specifically avoided. Such matters are relevant to a legal audit, or to the making of recommendations for improvement in the details of current operations; they are not significant when considering future needs in as rapidly changing and developing an environment as that in which the Company operates.

One important conclusion should be stated as the outset, however, and kept in mind throughout any review of these recommendations. The Company's legal function appears to be working well. There is no suggestion that this report was triggered by any really serious problem, and none is indicated. The many deficiencies complained of—overwork, inadequate information and controls, inability to "keep up" and the like—are not unusual in a company growing so rapidly, in an industry in which the regulatory atmosphere is tightening so much, and where so many different social trends are at focus on chemicals. Instead, such complaints may in fact reflect the common chemical-industry management philosophy that personnel should operate "lean and hungry."

Rather than evidencing dissatisfaction or inadequacy, the management decision to review long-term legal objectives, management structure and operations, reflects a willingness to assume leadership and control in an area which all too often receives inadequate management attention, to the detriment of all, including the legal profession. It is to be commended.

This report is lengthy, and may appear to some to propose a very major effort. In fact, however, what is proposed is in essence simple and often obvious. Legal terms and procedures are explained so that this report will be as useful to lay management as to attorneys; the result is that some recommendations (such as the use of stationery with a clear Legal Department marking, to help protect attorney-client privilege) may take longer to write about than to do. This need to communicate professional legal reasoning to lay persons, should not, by virtue of its length and apparent complexity, be permitted to obscure or interfere with implementation of relatively basic suggestions.

III

The Legal Responsibility

This section of the report will deal with the substantive legal responsibilities involved in the Company's legal function. The next section will deal with the matter of specific personnel requirements and assignments.

The only significant assumption of this section, is that there will not be any really major changes in the Company's business, or in government, such as might result from a chemical scientific "breakthrough," war, or national disaster. Other changes, such as those resulting from relatively normal growth, more or less government regulation and the like, may have an effect upon the personnel requirements discussed in the next section, but should not materially modify the essential legal responsibilities considered here.

1. *Preventive law*. The matter of "preventive law" is treated first because it is the one legal function which almost necessarily must be handled internally in a chemical company of any size. Outside counsel usually act only as "special" counsel, responding to questions or assignments as asked. Sometimes in organizations with fewer legal concerns, they will agree to undertake the "general counsel" function, and to assume responsibility that avoidable legal problems do not occur. But any outside assumption of such a responsibility here must be suspect.

Moreover, preventive law is usually the area which receives the least attention and shortest shrift in a growing company. By definition, it is not of the crisis or emergency character of so many other legal concerns. It can therefore always be deferred until a later time, when things quiet down (as they rarely do). In the long run, however, it is surely the most productive and profitable area. For what better measure of the success of the legal function, than that the Company has no unnecessary or unanticipated legal problems?

There seems general agreement among Company executives that the practice of preventive law is of major importance. One of the senior officers described it as "understanding the legal situation, and getting the organization into a position to comply." He attributed at least some part of the Company's success commercially to having anticipated the direction of governmental regulation and environmental concern, and in compliance developing new processes and products which would be superior to those of competition.

Discussion. In view of the size of the Company's internal legal staff, it is not surprising that preventive law is being practiced only in the most limited fashion. Corporate Counsel himself described his daily tasks as so overwhelming as to mandate that he almost always be occupied "putting out fires." He finds himself being able only to "react" to problems, rather than to plan and initiate. Educational programs and investigations of potential problem areas are rarely possible. Thus far, for example, he has been able to offer only one legal training session (in contracts.) No broad antitrust compliance effort has been introduced. Even the TSCA §8(e) compliance program remains to be fully implemented.

Many aspects of preventive law, which will be discussed in detail below, are really quite simple and obvious. They only require time and effort to administer. These include such matters as the enhancement of trade-secret protection; the maximization of attorney-client and work-product privilege; document retention policies; SEC compliance; indoctrination of personnel in antitrust, products liability, TSCA, adversarial relationships, and other legal areas; and centralization of claims and disputes administration so as to be able to pinpoint future problem areas and alert management to them. These aspects require only a designation of responsibility, and allocation of time and resources to carry them out.

Other aspects of preventive law, however, present problems and concerns which are among the most difficult any lawyer can face. Corporate "intent" and "purpose," for example, are frequently important concerns in the criminal as well as in the civil contexts. Just what is such "intent?" Was a marketing practice used to meet competition, or to gain an edge? Was a hiring or discharge practice based upon an intent to discriminate, or in order to carry out a reasonable policy of affirmative action? Even a Board of Directors mandate that there shall be no monopolization or restraint of trade, communicated to and signed by every employee in the Company, does not necessarily negate a corporate intent to do just the contrary.

Into this same area fall concerns with respect to "corporate credibility," "corporate responsibility," or "environmental awareness." Consistency between what is said and done by *all* persons in *all* areas of the Company is frequently critical; inconsistency with regard to such tenuous issues as corporate intent is the one readily available type of evidence which can destroy credibility, and impeach even the most persuasive executive.

The anticipation of new areas of legal concern and control is of the same character. Clearly, for example, the statutory, regulatory and common law relating to individual personal "privacy" is changing, and the Privacy Act, applicable presently primarily to government records, undoubtedly will have its counterpart in the private sector.

Record-keeping practices should be planned and controlled at their inception, to minimize future impact and cost, and to limit the potential for liability. Other such areas which should be the subject of "preventive law" consideration and report to management, include the matter of foreign "incentive" payments (or "bribery," depending on perspective), the installation of environmental controls in excess of those required by law, the doing of business in places such as South Africa or Chile, or the matter of outside directors' responsibilities and rights.

It should be apparent from this latter discussion that preventive law in these areas requires the involvement of the Company's General Counsel in all the intimate details of Company planning and operations. It also requires that he be in complete control of the legal function, and that all outsiders be responsible to him for what they say and do.

Recommendations. The following specific recommendations are designed to enhance the Company's preventive law effort.

a. *Internalize.* The Company's legal function should be internalized to the maximum reasonable extent. Where outside assistance is needed, tight quality control should be maintained.

This recommendation is designed to insure that the legal staff is privy to all major Company policy, planning, and operations, and is also in a position to be sure that what is said and done by outsiders on the Company's behalf is consistent and credible. It will also help insure that there is a funneling to the Legal Department of information concerning problem areas, so that red flags may be recognized and others alerted to potential problems.

b. *Team.* The "team" concept should be employed to deal with matters likely to involve major legal problems, so that all the relevant corporate and legal disciplines are brought to bear on the issues, with control and responsibility centralized. This recommendation is also designed to insure that the Legal Department is informed, and is in a position to participate actively. It enables it to offer legal advice even in areas which others may not always recognize as posing legal problems, and promotes consistency. Antitrust marketing problems, TSCA §8(e) issues, and claims/litigation review activities, are examples of matters which might be administered in this fashion.

c. *Education.* A continuing educational effort should be undertaken in the major problem areas, such as antitrust, products liability, environmental, OSHA, TSCA, SEC, EEO and the like. This should include written guidelines, disseminated to all employees and to new personnel as they are hired, and kept up to date. The guidelines should be supplemented by oral or videotape or videocassette presentations, as appropriate. and by special discussions on such matters as trade-secret protection, adversarial relations and the "legal" contrasted with the "scientific" and "commercial" methods (for dealing with government investigators, agents and other representatives.) A regular Legal Department newsletter should alert all personnel to matters of interest, such as proposed new regulations, legislation for which support is sought, new rules as to liability, or new judicial decisions of importance.

d. *Trade secrets.* The Company's technology is considered valuable, and a program for maximizing trade-secret protection should be documented and enforced. This should include not only the research-development records and registers already being used, but procedures for monitoring the handling of Company confidential information by licensees, for the limitation of access to secrets to those having a real "need to know." and for the classification and declassification of information (such as by marking it "confidential" and protecting it by lock and key, etc.)

e. *Privilege.* A program should also be developed for preserving attorney-client privilege and work-product exemption. Stationery used by attorneys should bear a "Legal Department" imprimatur. Attorneys should identify themselves as such, so that their status (as distinguished from that of nonlegal personnel in the Department) can later be recognized by those reviewing documents who have no personal knowledge of their positions. Letters and other materials should deal with single subjects, so that commercial and legal advice are not mixed and privilege thereby lost. Distribution of legal advice should only be to those in the chain of command. In contested matters, a "Re" caption should be employed, so as to identify work-product materials. Legal matters calling for interaction between nonlegal sections of the Company, should still be routed through, or at least so as to include the Legal Department, so that privilege is protected as much as can be.

f. *Record.* One of the most critical of Legal Department responsibilities, is the monitoring of the corporate "record" as it develops. It is a major duty of the corporate attorney to see that evidence is properly developed, to protect the Company and its executives against future liability, criminal as well as civil. Thus, if a lay executive requests legal advice, a memorandum should be made of the advice given, so that any future inquiry will at least establish that the executive sought to follow the law, and did not act "wilfully." Such a memorandum can spell the difference between criminal and civil liability, or personal and corporate liability, in future years when memories are dim and perhaps witnesses no longer available.

There should also be stated and enforced document-retention policies to insure that documents are properly preserved or eliminated in accordance with regular course of business practices and not in response to emergencies. The Legal Department should also monitor the preparation of, or at least the record of, significant corporate documents. The time to correct a mistake, whether in action or in simple expression, is when it occurs and *not* when a problem arises and the correction is viewed as simply "self-serving."

g. *Legal audit.* Finally, the Legal Department should also exercise an "audit" or "enforcement" function, much as the comptroller's office acts in financial matters. It should be on intimate terms with other Company operations as well as its own, so that it will be informed of developments which others might not recognize as presenting legal concerns. In certain areas, such as SEC, it should be furnished reports of compliance, so as to be in a position to make the necessary representations and give opinions as to validity and legality. In other areas, it should make spot checks (such as at trade association meetings), simply to monitor what is happening and have confidence that it is in accordance with representations.

One view was expressed that such an "enforcement" or "police" function was not appropriate in the Company—that a far better management philosophy is to trust people to do what is right. There are several reasons for suggesting a contrary course. (1) It is often difficult to tell what is right and what is wrong. The Supreme

Court's recent *Bakke* decision involved six separate legal opinions, and one is still not sure what the law is. Antitrust can be even more uncertain—the Company itself has experience in which different counsel have rendered different opinions to it. (2) Directors, senior executives, and management today are more and more frequently being held *personally* liable, criminally as well as civilly, for what is done by others in the organization. A Legal Department enforcement or "audit" program goes a long way to negate such responsibility by demonstrating that reasonable steps were taken to prevent the misconduct or error, and that the persons acting improperly were authorized only by virtue of their positions and not specifically by anyone else. (3) The law is beginning to impose a responsibility on senior management to police internal affairs, rejecting the contention that "I didn't know" if something could have been discovered by reasonable diligence. An example of this is the recent move toward requiring outside audit committees in certain corporations (although of course that move also has other conflicts-of-interest foundations.) (4) In any large and substantial company, it is almost a certainty that *some* representatives, agents or employees will go astray, no matter how carefully selected or how enlightened management may be. In one very recent litigation, for example, a senior partner of one of the major New York law firms admitted perjury and suppression of evidence, acting solely and exclusively on the client's behalf. Although there is reason to believe that the client was held responsible in a major way for the misconduct, it appears that it had no knowledge at all of what was going on. (5) Finally, there is a growing philosophy among some segments of our society that a person who makes crime or violation easy is as much responsible for what he loses as the actor—an employer leaving a wallet full of money on a bed, deserves the later theft by a domestic during the evening. Reliance in trust alone as a management tool may no longer be as valid as it once was. In this view, an enforcement effort is fair to the potential wrongdoer; without it, the company's rights may be limited.

2. *Regulatory law*. Although regulatory problems and concerns are critical to major segments of American business today, they are in fact a relatively recent arrival in the scientific and technical environmental-related areas. Air and water legislation of quite limited scope have been on the books for a generation and more, but OSHA, TSCA, FIFRA, CPSC, and the more aggressive enforcement activities of EPA must be considered creatures of the 1970s. It is not

generally appreciated that our institutions have yet to accommodate to these new realities. Law schools are only beginning to emphasize the regulatory area as basic; few businesses have identified and isolated the regulatory function (as distinguished from its specific components, such as OSHA or TSCA.)

Much of regulatory law can and should be handled by lay persons, just as they handle tax and SEC issues. But there is need to bring together the common regulatory issues so as to develop a uniform corporate approach. Because so many of those involved in these matters in government are attorneys, there is good reason to place the central responsibility in a Legal Department which is similarly made up largely of attorneys.

Wholly apart from improving compliance and avoiding inconsistencies, there is a very great deal that can be accomplished by this kind of measure. As one Company executive stated: "I will save a nightmare everytime I can educate a bureaucrat *before* legislation is passed or action taken." He quickly also pointed out, however, that "there are more issues than I can handle—the needs are far in excess of what we can afford."

Recommendation. A single senior Company attorney should be assigned responsibility to advise others in the Company with regard to issues relating to government regulation having a scientific or technical base. This would include TSCA; FIFRA; OSHA; EPA air and water; energy, F&D and CPSC if and when they become of concern; traffic and distribution; labeling; quality assurance; and government affairs and relations.

This attorney should have a scientific or technical background, so that he will be equipped also to handle patent and other legal matters having a significant technical base. At least over the near-term, it is hoped that most of the regulatory area will continue to be handled by Company nonlawyers, whose reception by government today is frequently superior to that accorded lawyers.

As will be pointed out in connection with the discussion of the responsibilities of the Associate General Counsel, regulatory affairs relating to financial matters (such as tax and SEC) should also be assigned single responsibility.

3. *Liability law.* This area includes responsibility to third persons and government in products liability, workmen's compensation, unemployment insurance, contract and regulatory violations, antitrust and SEC disputes, and other claims, contests, disputes, and litigations.

Discussion. Products liability may have burgeoned in other segments of American industry, but neither it nor the general litigation "explosion" appear to have posed serious problems for the Company. In part this may be because the Company does not concentrate on "consumer products," and is not labor intensive or faced with the problems of the typical service organization.

The "dispute-resolution" area, however, remains as one requiring attention. Many of the costs the Company presently bears are not always treated as legal costs, because matters are handled by outside counsel selected by third parties, such as an insurance carrier or an independent insurance adjustor, whose identity may not even be known to the Company's Legal Department. Perhaps more significant, there is no legal control over what is being said and done in such matters, because of the interposition of the third party between the Company and its legal representative, nor any mechanism for funneling back information so as to know what safety measures to introduce, what standards to modify, or what other corrective action might be appropriate. One Company executive stated that he believed there were a number of such cases, particularly in the workmen's compensation area, in which the Company should have its own legal counsel present, in addition to that of the carrier. Such duplicate representation may in some cases unfortunately be necessary, but it certainly represents a waste of resources, and is a poor way to proceed. It would be far better for the Company to handle its own representation, or to work out some new arrangement with its carriers to permit advance agreement on procedures.

Recommendation. All matters involving disputes should be handled by the Legal Department, which should itself select outside trial counsel when and if needed. This would include claims, collections, workmen's compensation, unemployment insurance, automobile and other vehicular negligence, products liability, and all other controversies in which the Company is involved. Where an insurance carrier is involved, procedures should be agreed upon with respect to the selection of outside counsel, where needed, and handling of the matter should be carefully monitored for the Company.

A claims-and-litigation review procedure should be formalized, so that each matter in dispute is evaluated, a strategy developed, exposure and target results estimated, cost and authorization controls introduced, and suggestions forwarded to management for correction, if appropriate.

A Claims Investigator should be appointed, with responsibility to investigate each accident or episode, as it occurs, or to select others to investigate where time or work load requires. To the extent available, this person would also be responsible for providing investigative assistance to counsel in other Company litigations.

4. *Other functions*. The following is a more abbreviated discussion of the other functions which are a part of the Company's legal responsibilities. Although also of critical importance to the Company, they are the more traditional areas in which only minor changes (other than internalization) are suggested.

a. *General commercial*. This includes purchases, sales, acquisitions, mergers, employment agreements, leases, licenses, uniform commercial code problems, and other commercial matters of all kinds.

There is a very considerable commercial advantage in being the first to draft a proposed agreement (for the other party has to justify changes, and cannot appreciate the inevitable "hidden" benefits), and this area of activity if fully internalized, might well take up the full time of a senior attorney and more.

It is recommended that an initial effort be undertaken to approve standard purchase and sales agreements which will require a minimum of change, and to develop "boilerplate" provisions for leases, licenses, employment agreements, and other contracts. Some of this is of course already being done, but it is hoped that by expanding the effort, reliance on outside counsel can be limited to very specific and discrete inquiries with regard to specialized problems, such as relating to "leveraged leases," and the general commercial area can be handled by a single Company attorney on something less than a full-time basis.

b. *Patent, trademark, copyright*. Although unanimous confidence was expressed in the Company's outside patent counsel, it is still believed that this function should also ultimately be internalized. (Trademark and copyright are joined with patent law here, because that is so commonly the practice.) Company internal patent counsel should in time be far more expert in the Company's special patent problems and concerns than any outsider, and should require less education by other Company personnel. He should be able to draft licenses for the use of technical Company information, participate in patent/antitrust decisions, and handle patent solicitations and contests, whether in the Patent Office, in court (assisted by trial counsel, if necessary) or on appeal. The Company should have its own docket, diary, and tickler system (described below), alerting it to the times for domestic and foreign filings and follow-ups.

c. *Tax*. There is every reason why the overwhelming part of the Company's tax problems can and should be handled by nonlawyers. But there is need for a Company attorney who is skilled in at least the usual kinds of tax problems the Company faces. Except on rare occasions, it should not be necessary to consult outside counsel or an outside accounting firm; when it is necessary, it should be done by internal tax counsel, who can hold the issue to the specific expertise required, limit the cost of the consultation, be himself trained in the process so as to be able to handle similar matters alone the next time, and insure high quality by bringing the Company perspective to bear throughout.

d. *SEC*. Most SEC problems also do not require legal involvement or legal advice. A Company attorney should specialize, however, in the SEC *legal* problems of the Company, and should be in a position to render whatever legal opinions may be required in the execution of agreements or otherwise. With exceptions, it should not be necessary to retain outside counsel simply to give an opinion as to legal validity or to pass upon problems which are likely to recur.

e. *Labor*. This area includes employee relations, employee benefits, EEO, workmen's compensation, and unemployment benefits (the dispute aspects of which are dealt with above and which should be handled by the same person who handles labor issues generally), and general management advice on labor matters. This also should be internalized, with outside counsel used only in very unusual circumstances.

5. *Administration*. Administration of the Legal Department is included here as a separate function, because it is so important, and yet so often inadequately considered—in private law firms as well as in corporate law departments. Lawyers are notably conservative, and are among the last to introduce modern procedures for central filing, word processing, text editing, electronic data retrieval, and similar aids to operations.

As suggested in earlier parts of this report, the Legal Department can and should be a vital aid to management planning, especially in a competitive and regulatory environment as rapidly changing as the Company's. The Legal Department is often the first to be in a position to discover outside perceptions and views which are in conflict with the Company's. Proper administration and information collection and control can help predict which products will pose problems and which will succeed, what the future regulatory atmosphere will be, and other such changes in the general commercial climate.

Administration of the Legal Department should be by a nonlawyer, of course under legal supervision. Lawyers do not usually like such assignments, and as a result may not handle them well. Moreover, legal salaries are frequently much higher than others, and can be justified only for strictly legal assignments. The "paralegal" can and should perform as much of the Legal Department's work as possible; paralegals are one of the most significant developments in years in terms of controlling cost and improving quality in the law.

The details of administration of the Legal Department are brought together and set forth below in the description of the work of the proposed Legal Administrator.

IV

The Legal Staff

This section of the report will consider the personnel required to handle the Company's reorganized legal function.

It was pointed out in the last section, that the nature of the Company's substantive legal responsibilities should not change materially over the next several years, absent some major change in the Company's business or in government. In contrast, the number of persons who will be required to handle the legal responsibilities of the Company, is a very direct function of the size and nature of the Company's business, government regulation and control, and the general economic and social climate.

It is believed that the four senior attorneys whose responsibilities are described below, supported by the Legal Administrator, the Claims Investigator and necessary secretarial help, should be sufficient to handle the Legal Department assignment. This belief is based upon the following assumptions.

The Company will achieve the annual growth predicted by management.

There will be no major changes in the Company's historical litigation pattern.

Government regulation in the air, water, energy, and toxic substances areas will continue to grow, although overall at a somewhat reduced rate from that of the past few years.

Tax, civil rights, labor, and SEC problems will remain fairly constant.

Antitrust problems will burgeon as the Company expands, in light of its major position in certain markets.

(There are some quite recent contrary indications that regulation and "big" government are coming into public disfavor, suggested, for example, by the success of Proposition 13 in California and the moves towards airline and communications deregulation. However, these trends do not seem yet to have touched the chemical industry.)

Changes in the above assumptions will affect staffing, but should not pose really serious concerns for the Company. The senior attorneys will all be in place, and ready to assume responsibility. The increased work load will mean only that more junior attorneys must be hired; some may be relatively young and inexperienced. It is recommended that such juniors not be employed until the senior staff is employed in any event, so that the matter of predicting Company and societal trends can properly be postponed until a time when more current information is at hand.

Some other preliminary comments are appropriate before turning to a description of specific staff assignments and responsibilities:

Synergism. Some of the functions in Section III might be handled elsewhere in the Company, and not by the Legal Department. Tax and patent matters, for example, are frequently administered by separate groups, outside of the Legal Department. Although there does not appear to be any justification for such separation except historical (and perhaps personal), it does not pose problems in the larger companies in which it is employed. However, it would be a mistake to permit such separation here, because it would deny the Legal Department the very important synergistic benefits of a larger staff:

Attorneys who are isolated in commercial organizations, frequently become more business- than law-oriented. Their intimate relationships are all commercial; the reward structure appears to be through general management; the best practice seems to be through "risk taking," rather than "risk assessing" and subsequent communication of risk to the nonlegal risk taker. This distorts the legal function.

Attorneys, as other professionals, thrive on exchange with colleagues. It sharpens the mind, keeps one up-to-date on current developments, and makes it possible to consider professional as well as mundane matters.

Law today is itself very much an interdisciplinary practice. Any worthwhile acquisition or merger, for example, requires expertise in a number of separate legal areas—tax, SEC, antitrust, patent, environmental, contract. Having these specialities all under one roof and responsibility makes it possible for the corporate law department to function just as comprehensively and effectively as the larger private law firm.

A 4–5 person law office should be large enough to enjoy many of the benefits of synergism. Larger offices are of course often necessary, but in fact often also pay a substantial price in administrative burden and loss of professionalism.

It is conceivable, although certainly not likely, that the four-person Legal Department legal staff recommended here will not always be occupied fully with legal matters. If and when that occurs, it is recommended that the staff remain as is, and that the attorneys be utilized in other nonlegal assignments suited to their training, such as contract or labor negotiation or government relations. It should be clear that these are part-time, ad hoc assignments created for administrative purposes, and *not* deprecation of, or diversion from, the underlying legal assignment.

Timetable. Lawyers are specialized professionals, whose personalities and work habits are not always predictable. The attorneys whose assignments are described in this report will all be holding relatively senior positions of responsibility in the Company. Their education, indoctrination, and development of relationships will take substantial time. It is not at all unlikely that one or two of them will not "fit" as completely as the assignment requires, and that they will leave or be given other assignments or released.

As earlier emphasized, no crisis or emergency is apparent. Accordingly, two recommendations are proposed with regard to timetable:

> The Company's long-term plan for the Legal Department should contemplate the hiring of only one senior counsel at a time, and anticipate a full year of experience before permanence is assumed.
>
> Operations with outside counsel should continue as at present, in parallel, until management is satisfied that each new staff counsel is performing his assignment satisfactorily, and on a permanent basis.

Hiring. Lawyers are usually ambitious, with substantial egos. The ranks of corporate law offices and private firms are full of attorneys who are anxious to move on, and who consider their careers blocked by senior counsel. Training is expensive and time consuming. It is recommended that an effort be made to employ counsel to fill the four new company senior staff attorney positions recommended in this report from the ranks of other chemical companies.

There are "headhunters" who specialize in legal positions, and who quite likely already have suitable candidates to suggest. Moreover, as in other cases, they can be used to approach anyone who appears suitable, without the embarassment which might result from a more direct approach to someone who might later be involved in a friendly trade-association activity, or in an adversarial commercial transaction.

It is also recommended that Company executives consider the qualifications of attorneys from the other chemical companies with whom they have worked in industry trade associations and elsewhere, and suggest names of probable candidates to the recruiter.

A minimum of perhaps three-years chemical-industry experience is needed; more than ten years would probably place any attorney out of price reach (or if not, suggest a problem in personality, work habits or otherwise.) Provided that family and other personal considerations do not stand in the way, it is a good bet that a suitable approach to such an attorney will be productive.

Compensation. Legal salaries are today very high. Some of the major law firms in New York City are offering $37,000 first year incomes to the top law-school graduates, and on up to $40,000 and above, including additional fringe benefits, such as car and apartment, for special assignments (as in the Government's antitrust case against IBM); 10–12% annual increases are in order for many. Salaries in other cities, or in smaller law firms for candidates with lesser qualifications, are scaled down from these levels, but they are all very likely to be considerably higher than for most other professionals or for lay persons.

It is recommended that the Company have a separate compensation program for legal professionals, competitive with that of other chemical companies. Employment of legal professionals should also be free of the other regular constraints of the Company's personnel practices, to the extent that competition requires.

Assignments. The four-person legal structure recommended, and the specific assignments to each attorney, are somewhat unusual. Ordinarily, for example, the tax, patent, antitrust, labor, and trial specialties are handled by separate attorneys, who pretty much each stick to their own lasts. The size of the Company's legal function here, however, is such as to mandate a departure from usual practice.

The recommended assignments are along logical substantive lines. Obviously some changes may be appropriate, to accommodate the personal predilections and backgrounds of particular attorneys as they are hired. Clearly, also, it will be necessary for attorneys to cross assignment lines and assist each other to help smooth out the inevitable peaks and valleys of work load. An effort should be made, however, to keep to the recommendations in general; over the long-term, they should be more workable than deviations suited to special requests or needs.

Titles. Titles should not be important, of course, but in the legal profession at least, they are. Outsiders with whom staff counsel deal will draw conclusions about the nature of an attorney's responsibilities and authority from the title alone, and often will not inquire further. The titles "General Counsel," "Tax Counsel," and "Patent Counsel," for example, may be openings into such organizations as the "Association of General Counsel," or leadership in trade-association activities.

It is recommended that the titles of "General Counsel," and "Associate" or "Deputy" and "Assistant" General Counsel be used. At the same time, however, it should be very clear on hiring and throughout, that titles and organizational chart assignments are descriptive only, and that the actual work assignment of any staff attorney is subject to the direction of the General Counsel. Attorneys sometimes develop strong convictions about their rights and responsibilities; the proposed Legal Department is much too small to permit anyone to concentrate exclusively on his areas of special competence or responsibility at all times. Everyone must be available to jump into any breach requiring a major effort, such as a disaster or work stoppage; each must be available to help smooth out the work load, as above, to permit feasible vacation scheduling, and to respond to emergency.

Expansion. If the rest of the Company finds the new structure useful, there may well be some tendency to transfer quasi-legal functions (such as tax, government relations, SEC) entirely to the Legal Department; unless there is restraint, the Legal Department may itself find this attractive and will be amenable to expansion.

It is recommended that there be a conscious and continuing effort to continue to handle "quasi-legal" matters outside of the Legal Department, where possible. Lawyers are (or should be) intelligent, well-trained people and, given the opportunity, can "take over" more and more of the commercial function. This should not be permitted to happen. Lawyers should be risk assessors and advisors; risk taking and operating responsibility should be the province of business people. Business today is as much a specialty as law, and can better be handled by those expert in its intricacies. Moreover, the legal profession has lost much in credibility, and in many cases a scientist or production expert can far more effectively convince a government representative (or customer) than a lawyer.

Much of government regulation, tax, contract drafting and other day-to-day commercial activities, should therefore remain as is. At some point, however, and despite all efforts to resist growth, it will probably be necessary to hire new attorneys in addition to those specifically identified in this report. This may be because of workload requirements, for succession purposes, or because of new developments.

Very large private law firms and corporate law offices like to hire direct out of law school. It is recommended here, however, that future Company junior staff attorneys have a minimum of two years experience elsewhere, preferably in the chemical industry (and not as a judge's law clerk, or in a prosecutor's office, for example). The Company's expanded Legal Department will still not be really large enough to make possible a sound basic training program; attorneys are frequently disappointed at what they find to be the reality of law practice, in contrast to what they thought it would be in law school—they blame the inadequacies on their first job, with resulting high turnover at that time; training and education is a very expensive assignment, which might well interfere with the regular operation of such a small Department, and certainly could pose problems for nonlegal executives who want legal advice rather than teaching experience.

Although the staff described below should be sufficient to handle the Company's legal matters, as already stated and on the assumptions set forth, it is deliberately a minimal or "bare-bones" staff. Despite constraints, the resulting pressures may compel expansion if the activity is as successful as hoped. This need not be a sign that the legal function has gotten out of hand and that the lawyers have "taken over." It is hoped that instead, it will be evidence that the Department is doing its job well.

1. *General Counsel*. The General Counsel is the Company's chief legal officer, and responsible for the legal function. He should be kept informed of all major policy and operational matters, and should generally participate in important discussions of problems likely to involve legal concerns.

Although there is some contrary opinion in the private bar, the modern and better view is that the General Counsel should not be a Director of the Company. Certainly he should sit with the Board, and be privy to everything which a Director would consider. But it is worth repeating that risk assessing and risk taking are two very different functions. As one senior Company executive stated, "I tell my lawyer not to practice business, and I promise not to practice law." Directors of course *must* administer the business and decide on what risks to take; a lawyer-director cannot escape such a responsibility. Yet it is a very rare lawyer indeed who can present an impartial risk-assessment to the Board, while he is also arguing in favor of a course of action involving that very risk.

The General Counsel should of course administer and manage the Legal Department. All outside counsel should be appointed by and report to or through him. Only in the most exceptional circumstances should outside counsel consult with or advise the Board or senior management in his absence. Any contrary course negates his authority and responsibility, and deprecates his position.

The General Counsel also advises the Board and all senior Company executives. But in this latter connection, one caveat should be stated: More and more problems today involve a potential for personal executive liability or responsibility. The Justice Department has already announced its conviction in such circumstances that corporate counsel who represents the executive and his company is in a conflict, that privilege is thereby lost, and that the conflict itself constitutes evidence of impropriety. It should be made clear to all Company executives that they are entitled to personal counsel of their own choosing whenever such a problem arises, and that the decision to so proceed will not under any circumstances be viewed as evidencing disloyalty or impropriety.

All staff attorneys of course report to and through the General Counsel. In addition, because the proposed Legal Department is small and cannot afford the luxury of a single attorney devoted to administration and senior executive advice alone, the General Counsel should personally assume responsibility for the following substantive legal areas:

Antitrust, including trade-association activities.
Educational programs and legal guidelines.
Major contracts, acquisitions, and mergers.
Product stewardship and corporate responsibility.
Special foreign-law problems.
All matters involving potential criminal responsibility.
All matters involving the personal liability of senior Company executives and Directors.

As with all other staff attorneys, the General Counsel should be ready to fill in and assist other attorneys, as required.

2. *Associate* [*Deputy*] *General Counsel* (*Commercial*). The Associate or Deputy General Counsel is the number-two person in the Legal Department, and acts for the General Counsel in the latter's unavailability. He may also be called "Tax Counsel." The descriptive word "Commercial" (or "Financial", which equally describes his substantive assignment) should be used for internal purposes only; the title "Associate General Counsel" is sufficient to describe his authority to outsiders. Anything further would be interpreted as a limitation or restriction, and as distinguishing him from one or more other Associate Counsel. (The word "Deputy" is often used rather than "Associate," and is equally acceptable.)

The Associate General Counsel should have responsibility for the following areas:

Contracts of all kinds, including sales, purchases, acquisitions, mergers, employment and licenses—but not ones dealing with patents and technology or labor relations.

Tax.

SEC, FCPA and related accounting problems.

Real estate.

ERISA.

Corporate records and minutes. (It might be useful for him to also be an Assistant Secretary of the Company.)

Other government and regulatory matters with a financial or commercial base.

Miscellaneous other commercial matters.

3. *Assistant General Counsel—Patent.* This attorney is also given the title "Patent Counsel," because it is a traditional term which will permit him to work with the Patent Office as well as with patent-attorney colleagues and adversaries in optimum fashion. However, his responsibilities might more accurately be described by using the word "Technical," or perhaps "Regulation", for his assignment is to handle all Company legal matters which call for significant technical or scientific expertise (except judicial litigations, which either he or the Assistant General Counsel—Litigation might handle, depending upon the specific circumstances in each case).

Company Patent Counsel should handle the following areas:

Patent, trademark and copyright matters, including contracts and licenses in these areas.

All EPA matters, including TSCA, FIFRA, air and water pollution.

All OSHA matters.

Energy, F&D, and CPSC, when, as and if they become problem areas.

All other environmental matters, including state and private.

Government relations and affairs, state, local and federal, including legislative and administrative advice and comment on proposed regulations.

Traffic, transportation and distribution.

Safety.

Quality assurance.

Labeling.

Packaging and packaging standards.

Industrial hygiene.

Other government and regulatory matters with a technical base.
Miscellaneous other technical matters.

He should review the *Federal Register* every day, and advise others (legal and nonlegal) of pertinent matters, even though many of the others should read the *Register* themselves. He should extract important items for publication in the Legal Department's regular newsletter. He should be informed of all major Company relations with Congress and the executive.

4. *Assistant General Counsel—Litigation.* This Assistant's title (which might also be "Trial Counsel") would more descriptively be "Dispute Handling," for only a small part of his activities will be dealing with litigations in the sense of judicial contests in court. His responsibilities should be as follows:

Administrative and judicial litigations.
Arbitrations.
Claims by or against the Company.
Collections.
EEO.
Employee benefits.
Employee relations.
Insurance, including unemployment insurance claims.
Labor issues of all other kinds.
Products liability.
Vehicular negligence.
Workmen's compensation.
Miscellaneous other dispute matters of all kinds, including all hearings, trials and other legal proceedings (excepting limited rule-making and similar administrative proceedings, and other matters already assigned elsewhere.)

Assisted by the Legal Administrator, he should handle the claims and litigation review function. He should try arbitrations and administrative cases, such as EPA or workmen's compensation hearings. He should also try judicial litigations to the extent of his competence.

Where the Company continues to work with outside counsel selected by an adjustment bureau or a workmen's compensation insurance carrier, Litigation Counsel should monitor the representation carefully, and be prepared to step in as required. It is worth emphasis that at present there is no significant Company involvement in the legal aspects of these matters, and that attorneys in distant places are speaking for the Company without legal restrictions on their authorities. The inadequacy of information and risks are apparent.

a. *Claims Investigator*. A person trained in investigative matters should assist the Litigation Counsel in all matters requiring factual investigation. This would include accidents of all kinds, claims, litigations, collections, product defects, personnel misconduct, trade secret violations and miscellaneous similar problems. If time permits, this person might also handle other Company investigations, such as employee applications where special security matters may be involved or unique market conditions.
Recent cutbacks in products liability and other insurance coverage have limited the advancement potential and possibilities for many first-rate insurance company senior claims investigators. It should be possible to employ such a person from one of the insurance companies, including even one which has been serving the Company in the past.

5. *Legal Administrator*. It is the primary assignment of the Legal Administrator to permit the licensed professionals to practice law, and not to spend unnecessary professional time on matters which a lay person can handle. This includes a host of unrelated assignments, such as identifying costs, marshalling information so that the Department can perform its early warning role, and processing educational and document control matters.

The importance of this position is frequently deprecated by the attorneys, as earlier stated, and therefore deserves emphasis here. Without it, filing dates may be missed, statutes of limitations overlooked, and control lost, unless the attorneys fill the gap. It is worth repeating that the attorneys would not perform as well; certainly the out-of-pocket and other costs would be far greater.

The assignments of the Legal Administrator are as follows:

Docket, diary and tickler. A regular docket and diary should be maintained, listing every legal event, with a tickler system notifying the appropriate person of the need for action at the right time. This would include follow-ups on patents, trademarks, copyrights, corporate filings, contract renewal and performance dates, litigation filings and other actions, statutes of limitations, regulation comment dates, licensee commitments (such as for maintaining the confidentiality of trade secrets), and a host of other events far too numerous and broad to be able to anticipate here. It would also include outside litigations, for every so often an outside attorney will slip up, or there may be differences of view and even a need for change.

Every attorney in the Department may well keep his own record as a check, but there should be confidence that the Legal Administrator will alert the proper person when something must be done, in sufficient time to permit action to be taken, and even if someone is sick, in a hospital or otherwise unavailable. There is far too much at stake to permit long-run reliance on the fallibilities of human health and memory.

Trade secret administration. A central record should be maintained of Company patents and technology, keyed to development and invention data to establish ownership, and to security controls to establish confidentiality. Regular course of business procedures should be included to insure that access to confidential information is restricted to those having a "need to know," in the hands of licensees as well as Company personnel; that precautions as to security are properly maintained; and that information is properly classified *and* declassified. (Declassification is important evidence of a good trade secret program. It helps prove that still-classified data is really a trade secret. Improper continued classification, in contrast, can prejudice a trade secret program. Maintaining lockbox security and claiming secrecy with regard to something clearly in the public domain, for example, can throw doubt on the balance of the security program.)

Attorney-client and work-product privilege. The Legal Administrator should also be responsible for insuring that guidelines with regard to maintaining these areas of privilege are being properly followed. Some attorneys like to broadcast copies of their memoranda widely to Company executives; such a practice will jeopardize attorney-client privilege, which requires that legal advice be disseminated only to those in the corporate chain of command who have a need to know. Legal Department stationery should always be used when legal advice is given, and attorneys should use some identification (such as their titles) to indicate that they are acting as attorneys. Commercial and legal advice should be in separate documents.

Regular business practice. Although one of the exceptions to the hearsay rule (which renders evidence inadmissible in judicial proceedings) relates to documents kept in the regular course of business, it is surprising how few attorneys actually maintain their own records in good "course of business" fashion. It should be the responsibility of the Legal Administrator to insure that such practices are followed—that, for example, documents are always routinely stamped with a time and date when they enter the Legal Department, so that there will never be a doubt what was received and when; copies should be routed in accordance with a regular and routine practice, so that who did and who did not see something can be established; files should be maintained so that something missing can be shown to have been taken (or not received), and not simply be the subject of misfiling.

Retrieval. The Legal Administrator should also handle the filing and retrieval system. This would include the paper files, maintained in accordance with some standard practice so that one attorney can take over another's files where necessary. It also includes computer retrieval of important documents, such as legal memoranda prepared by outside law firms (or by the Department itself), to avoid doing the same job two or more times; or comments by Company executives to government agencies, to avoid inconsistencies, which are always at the core of cross-examination and of impeachment of credibility. Such a retrieval mechanism will grow more and more useful as the data base is enlarged over the years, to the point where nonlegal personnel will come to rely on it in legislative testimony and public statements, and where it will be essential for witness preparation if, as and when there are litigated problems or government investigations. The time to begin is early, before the data has grown to proportions too large to handle easily.

Records retention. The Legal Administrator should also administer the Company's records retention policies, to be sure that what should be preserved is kept, and that what should be eliminated is properly disposed of. Although theoretically this should be a self-enforcing program carried out by each of those who keep files, in practice it never is. People need to be regularly reminded of the rules, on a companywide basis.

Claims and litigation control. Although the Assistant General Counsel—Litigation is responsible for claims and litigation review, he will need the active assistance of the Legal Administrator in handling this assignment. Implementation of the program is a detail of Legal Department administration, but as an illustration, one set of forms for litigation review is annexed as Exhibits 1A, 1B, and 1C. These include identification of responsibility for handling the matter, exposure and target results, plan and strategy of attack or defense, nature and extent of authority, timetable, and all other relevant concerns. All should be on computer, so that problem areas can be analyzed by product, geographic area, compound involved, individual, or what-have-you. Similar forms should be maintained for claims which have been asserted, and even for episodes (such as fires, accidents or product manufacturing defects) which are likely to produce disputes but have not yet matured into claims.

Retainer control. It is of course the function of the General Counsel to appoint and deal with outside counsel on substantive matters. But it is the responsibility of the Legal Administrator to follow up to be sure that outside counsel is complying with the terms of the retainer understanding. Exhibit 1C, for example, is the kind of regular litigation report which should be received from outside counsel, or which the Assistant General Counsel—Litigation should prepare on the basis of information furnished by such counsel. The Legal Administrator should see that the report is recieved when due, and then process and analyze the data to determine, for example, whether copies of legal memoranda prepared by outside counsel have been forwarded along or whether new attorneys have been listed as working on a matter where the agreement was to use others who had already been trained in the Company's affairs.

Any good private law firm devotes substantial management expertise to the analysis of records such as these, to determine who is efficient and who is not, where time was wasted, where activities were productive, and a host of other matters. Certainly a client who is paying the bills should do the same in substantial matters, where an independent perception is relevant. Legal services today are sufficiently expensive and important to warrant the attention which any company pays to other major cost-plus purchases. The limited effort should pay substantial dividends in quality as well as in cost reductions.

Attorney time records. The Legal Department's only product is legal services, and time is the best measure of such services. Each attorney should keep records of professional time devoted to each matter on which he works, with a dollar charge ascribed to that time that reflects true cost (including overhead). This is useful both for general management purposes (in permitting the same kinds of cost assessments as undoubtedly are employed in other departments of the Company), and for internal Legal Department control. It will show which attorneys are efficient and which not; where time has been spent without significant benefit; where time has *not* been spent, to disadvantage; and a variety of other information which will permit intelligent decisions. Attorneys themselves who have never kept time records, frequently find that the disclosures help them to better prioritize their time, and become more efficient.

It should be added that *all* Company costs attributable to legal matters should be collected by the Legal Administrator, including executive time spent educating attorneys or traveling to hearings, disbursements for transcripts, and charges for outside legal opinions. This marshaling of information will sometimes suggest product activities which might better be eliminated, or contests where discretion and settlement should be the better part of valor. The Administrator should be able to respond promptly to management requests for information regarding legal costs, reserves for claims, work loads and the like.

Legal research. The Administrator should handle all library purchases (which can become quite expensive), and the distribution and filing of current pocket parts and current releases. These can otherwise easily be lost in someone's home or desk, and the next party kept in ignorance of an important development. He should also explore, when appropriate, the possibility of introducing computerized legal research (such as is now offered by LEXIS or Westlaw). He should maintain contact with outside specialized legal research services, which may assist Legal Department staff. Such services are sometimes used by outside private law firms, with a markup in cost to the client on occasion where the research has been adopted as the firm's own.

Secretarial services. The Administrator should supervise the availability of secretarial services to the legal staff. The use of word processing and text editing equipment should be expanded to its practical maximum. Enormous amounts of standard paragraphs and provisions can be kept in computer memory, resulting in substantial savings of legal and secretarial time, and minimizing the necessity for proofreading.

Internal newsletter. The Administrator should edit an internal newsletter for review and revision by the General Counsel, and distribution to the Legal Department and other appropriate Company personnel. This newsletter would contain important general diary entries of which personnel should be informed, important developments of legal significance, and discussion of important legal issues (such as in antitrust or TSCA) which should be communicated as a part of the Legal Department's general educational function.

Paralegal services. One of the recommendations of this report, is that outside counsel handle only those matters requiring their special legal expertise, and that paralegal functions be either handled by the Company internally, or purchased by it directly. An extreme but most illustrative example is the computerized litigation support system which is required for the handling of any major litigation. When handled by outside firms, paralegals are hired to deal with coding matters, and frequently charged for at $30–$35 per hour and more; the same work can be done by the Company, hiring special employees at perhaps $6–$7 per hour, or by employing computer services firms at much lower than law-firm rate.

Professional legal services require a great variety of other support services, including printing, proofreading, duplication, messenger, investigative, and so on. Some of this can be quite expensive, depending on the matter—printing of certain financial documents needed for SEC purposes can cost several times as much in a large city as in the company's headquarter. All of these kinds of purchasing matters should be handled by the Legal Administrator.

6. *Outside counsel.* It is a major recommendation of this report that outside counsel should be used, over the long run, only for specialized and nonrecurring matters, and then only for the specific functions which the Company itself is not able to handle. Even in the long run, however, there will rarely be a time when the Company will not need to rely on outside counsel in some matters and to some extent (such as acting as "local counsel" in out-of-state litigations, where local court rule usually mandates the use of a local attorney).

Outside counsel should always be appointed by the General Counsel, as stated elsewhere at several places in this report, and should report to and be responsible to and through him, except in very unusual circumstances. Any other course will inevitably demean and deprecate the function of General Counsel, interfere with his responsibility, and bring into play a host of competitive conflicts between professionals, from which the Company can only suffer.

The use of outside counsel should be restricted to the specific needs for which they were selected. If some specific area of tax or corporate law is involved, for example (as in the so-called "leveraged lease" referred to earlier, and considered by one Company executive as calling for outside expertise), staff counsel should work closely with such counsel, and handle all the other aspects of the assignment, including the drafting and even the typing (which some firms charge for as a disbursement, at substantial rates), the proofreading, the filing, etc.

It is difficult for lay persons to appreciate how very much of the work of a private law firm is handled by nonlegal personnel, paralegals, or junior attorneys fresh out of law school, who may not have a fraction of the competence of corporate internal counsel. Even senior executives assume that a law "firm" is experienced and specially talented, simply because the senior partner with whom they work is. Frequently that is not the case, and the client's personnel also can handle the major part of the assignment. This has the additional benefit, of course, of training Company personnel in the particular expertise required.

There is inevitably resistance to this kind of working relationship with a private firm. Law firms do not make their profits on senior partners' time, for the senior usually earns far more than his billable hours can bring in. (A senior partner in a large city may bill $175–$300 per hour, and work about 1,200 billable hours, with the rest of his time being client development and *pro bono* work; overhead may be as high as 50 percent of gross income, and his share $150,000–$350,000. There is little or no profit here for others.) The greatest profits are made on paralegals, as above, or very recent law school graduates, who may be able to bill as much as 1,800 and more hours, at $50–$60 per hour and above. The ability to develop the necessary relationship with the outside firm will depend in very large measure upon the capacity of the General Counsel to negotiate in professional fashion, and to work out a retainer at the inception which is satisfactory in all respects. It can be done, but is obviously a most sensitive assignment.

This area has been discussed at some length, because it appears that present Company practices—similar to those of many others—permit the appointment and retention of counsel quite liberally, including even by local plant managers and outsiders, and usually without any written retainer agreements or even retainer understandings. If one includes collections, products liability litigation counsel retained by a carrier, and workmen's compensation cases, the Company probably has at least two dozen law firms working for it, the identities of some of which are unknown to its own internal counsel.

A review of recent legal bills indicated only a very few in which the hours worked were listed, and none in which individual, time and rate were specified. Most bills simply set forth professional charges in very gross and clearly "rounded" terms, sometimes with a specification of what was done in general fashion—such as "conferences with client," "consideration of law," etc. This was in the most marked contrast with the billings for disbursements, which were so detailed as to be almost ridiculous, and yet were for minimal amounts. One classic bill included a charge for 80¢ for the duplication of four documents, and then attached a duplicate of the invoice for the duplication job (apparently the firm absorbed the 20¢ charge for duplicating the duplication invoice).

One of the bills notified a senior Company executive that certain legal time was being billed at in excess of the regular rates, because the matter involved was a special one. Yet in fact, the only justification for regular rates of outside counsel is because of competence; a $175 per hour rate is fixed for an attorney because it is presumptively justified by his background and experience. To charge a second, higher rate should be permitted only where there is a specific understanding to that effect. This bill also confirms the need at the inception to be sure that the retainer is on a most-favored-nation basis (see Exhibit 2, item 4b), because so many firms do have more than a single hourly rate for each attorney.

Until such time as the appointment and reporting of outside counsel is administered through the General Counsel, it should be clear Company policy that the basis of any retainer must be agreed upon in advance and, except in unusual circumstances, reduced to writing in precisely the same fashion as any other purchase of services by the Company. Exhibit 2 attached is a retainer checklist, identifying the kinds of matters which should be considered, depending of course on the nature of the assignment, the Company's prior experience with the firm and all the other relevant considerations. In addition to most-favored-nation treatment, assurances should be received that attorneys trained in the Company's business will stay with the Company's matters in the future, assuming that they are acceptable to it. Authorization should be limited to specific matters, estimates should be requested before certain levels of charge are exceeded, and regular time and action reports should be furnished. Firms will usually be agreeable to the reasonable requests of a substantial client, and will themselves propose special provisions, such as "lids" or "caps" on fees, reductions in charges after certain levels have been reached, deferrals (which are sometimes most welcome for tax purposes), and client participation to limit work requirements of outside counsel.

It is not the intention of this section to deprecate the services of outside counsel, who can and do perform a most important function, and may on occasion be vital to a company's success and even survival. But, as the United States Supreme Court has confirmed, the law is a business as well as a profession. In light of problems with the outside bar referred to by the Chief Justice and others, in the present chemical industry environment (with legal concerns of such major importance and cost), there is every reason why the acquisition of legal services should be accompanied by the same quality and cost control measures so effectively applied elsewhere.

7. *Costs*. It is a conviction of this report that implementation of these recommendations will result in very substantial dollar cost savings to the Company, over the long run, considered in terms of the quality of legal services which will be provided. Unfortunately, however, it is not possible to document or prove this assertion in specific financial terms. Indeed, over the short term, as pointed out at the outset of this report (see Section I(7), *Costs*), indoctrination expense and parallel operations will undoubtedly result in some increased costs; over the long run, if the recommendations are sound, the legal function will grow so that no dollar savings will ever be apparent. Perhaps most important, there simply is no way of proving what cost savings have resulted from avoiding legal problems or reducing settlement outlays.

Unlike many other management ratios, there is no useful chemical industry figure to suggest what the cost of legal expense should be (such as some percentage of sales). There is far too much variation between those marketing consumer and nonconsumer products. One major antitrust or environmental litigation can throw a ratio completely off for many years. Different managements also have different concepts of how to use the legal function. And companies all appear to classify "legal" differently. Some include tax and/or patent in "legal"; others do not. Some include paralegals for some purposes and not for others.

Having said all this, however, it does appear that the average of $500,000 in Company annual legal expense during the past three years is on the low side, reflecting the legal activity of a somewhat smaller company. There should not be major concern if the planned reorganized legal function operates at that continued level and properly performs its assignment. It is recommended that this be the preliminary target, subject to early revision.

Ultimately the costs of claims paid and recovered should be included in any analysis. They may not be a perfect test, but they are the best available indication of the success or failure of any sound preventive law program.

It is believed that the reorganized legal function will provide legal services far more efficiently than at present. One of the greatest savings will be in the time of Company executives, who will no longer have to devote so much time to educating attorneys (or themselves) in legal matters. An executive stated, "If we took the money we waste educating outside attorneys and hired a legal staff . . . "

One indication of success will be when someone says "What a pleasure it is to just telephone our inside counsel, who knows just what I'm talking about, and just what I want," and, in glowing terms and with pride, tells others "*My* lawyer says . . . "

V

Conclusion

These recommendations are based in part upon conversations with Corporate Counsel and other senior Company personnel. The underlying predicate of all of the recommendations, however, is also based in important part upon a number of economic, social, and political factors unrelated to the experience of any single company. That predicate is that "internalization" of the legal function should be a long-range goal of any substantial company in the chemical industry today.

"Internalization" means that all legal matters should be handled, or strictly administered, by attorneys who are full-time corporate employees, with commercial loyalties exclusively to the Company.

Recurring matters, no matter how specialized, should be handled by staff attorneys, alone.

Where exclusive internal handling is impracticable, either because a specific matter requires a specialized talent not otherwise sufficiently useful to justify developing the needed expertise internally, or involves a major special staffing effort for which Company personnel are not available, tight control by internal attorneys is mandated.

The following are the most important reasons for internalization:

Increasing legal problems. The number and scope of chemical industry legal problems is huge, and seems to be growing at almost an exponential rate. Undoubtedly this is to some extent a reflection of the civil rights, social welfare, consumerism, environmentalism, and populism movements of the post-World War II years, fed by concerns about overpopulation, depletion of natural resources, and some public disenchantment with the corporate structure and corporate credibility. For whatever reason, however, the chemical industry today is in significant part a regulated industry—indeed, its problems in some respects are greater than those of the traditional regulated industry, such as the electric utility, because it is not guaranteed a return on capital and cannot set its prices by administrative filing and mandate.

Difficulty of legal problems. The difficulty and complexity of current legal problems has also grown, to the point where the quality of modern legal services is dependent upon an intimate knowledge of a great number of facts, including esoteric sciences, attitudes of company executives, policies as to corporate responsibility and the like. It is a major assignment to educate a few staff attorneys in all this; it is impractical and unsound to educate a host of different outside counsel on an almost continuing basis.

Skyrocketing external legal costs. The cost of privately furnished legal services has already been adverted to; it has for some years been almost literally skyrocketing (although there are indications that this is now coming to an end). Undoubtedly this is a short-term maladjustment, which will be corrected as more and more public and private organizations internalize (individuals cannot pay these charges), as the law schools turn out additional attorneys (until recently, admissions had been burgeoning), and as more and more competition sets in by virtue of recent judicial decisions. But one cannot expect significant real change for a generation, until the new crop of recent graduates becomes experienced and competent to take over major legal matters.

Inefficiency of external legal services. It is not critical of the providers of outside legal services to say that, almost by definition, they cannot be as effective or efficient as an internal staff.

They *are* outsiders, with their own businesses and careers at stake. They are necessarily concerned with the extent of their own in-depth knowledge of the client's affairs, and therefore far more conservative than would otherwise be.

Legal services are their businesses, to be expanded and enlarged. They are not a cost of doing business, as they are for the client, to be controlled and limited.

There is growing recognition even in the outside bar that law firms must do things for their own individual and organizational reputations and businesses, and apart from the client's purposes and objectives. A law firm's rotation program, for example, may be properly designed to train the young attorney in a number of different disciplines so that, for his own and the firm's good, he will become an all-around effective lawyer. He may therefore be transferred to another department or matter, even though he might also be just the person to work on a specific matter for a client by virtue of his previous client contact or experience.

There was a time when corporate inside counsel were considered and treated as "second class citizens" in the legal profession, so that corporate executives believed that they had to go "outside" for quality, or at least had to check the internal advice given with outside counsel. This may have been partly the result of an NIH (Not Invented Here) syndrome, as one Company executive stated. Yet it was also somewhat of a self-fulfilling prophecy, because inside counsel were hired at low salaries, were excluded from major legal problems, and did not always grow and mature as a result. But this is no longer true. Corporate legal counsel today are generally far ahead. They are on the firing line "where it's at," dealing with regulatory and administrative problems long before they ever reach the courts and become known to the outside bar and the law schools. Corporate legal counsel are *making* the law regarding TSCA—law with which judges, professors and outside attorneys may not be dealing for some years. The same is true in a great number of other areas, from water and air environmental law to the problems of employee privacy, dealing with corrupt or hostile foreign governments or employees, or "reverse discrimination."

It may be true that an outside patent attorney, by virtue of years of concentrated and general patent experience, will have more knowledge of the general patent field than an internal staff attorney. But the significant corporate need is not for such generalized competence. It is for very detailed knowledge of the Company's specific patents and patent problems. No one could possibly be more skilled and expert in that area than a staff attorney who has devoted all of his or her career time to acquiring the needed factual information, studying the particular scientific specialty or discipline involved, working (and having regular luncheons) with the scientists involved, and understanding the organization's objectives and motivations. Precisely the same is true of other specialties, whether involving problems of environmental pollution, antitrust economics, employment opportunity, or otherwise.

The Company staff attorney should be like the top-notch Company accountant or occupational physician. There may be outsiders with greater expertise in general areas but none with greater expertise in the specific concerns of the Company and its industry.

The Company effort which has gone into the development of this report has been substantial. Undoubtedly Company executives will devote considerable additional effort to considering its recommendations. Such attention to the legal function is unusual, and perhaps unprecedented.

Continuation of such attention is unlikely. Rather than let developments just happen in the future, it is urged that these recommendations, as modified by executive action, be used as a guide to future Legal Department action. They may be changed, of course, but only by advertence and decision.

The long-term goal of the Company should be to develop an internal legal staff which is more skilled in the Company's specific and recurring legal problems than any other attorneys in the world. Unlike many other chemical companies, it has a chance to build afresh and anew, with a view to accommodating to the new legal rules of today. The structure of the chemical industry is such that the majors are frequently concerned with leadership, and are anxious to defer to smaller companies with expertise, in trade associations and elsewhere. Leadership in this area by the Company is not only a societal benefit, but will help it fashion its industry and the law to its own benefit.

Exhibit 1A

Litigation Review Form

1. Name of case:		
2. Court:	—Brief description:	
3. Claim:	—Monetary amount:	
	—Other:	
	—Insured?	
4. Counsel:	—Firm:	
	—Individual in charge:	
	—Corporate attorney assigned:	
5. Fee:	—Written (attach copy to individual report).	
	—Oral (describe arrangement):	
6. Litigation plan:		
	—Written (attached copy to initial report and when revised).	
	—Oral (describe, including anticipated trial date):	
7. Evaluation:	—Maximum realistic exposure	—Monetary:
		—Other:
	—Target result	—Monetary:
		—Other:
	—Anticipated litigation costs:	
8. Comment:	—Does trial counsel advise currently?	
	—Settlement appraisal?	
	—Other?	

Exhibit 1B

Litigation Preparation Plan

(To be revised as appropriate, and kept current)

Anticipated Timetable for Commencement and/or completion

I. Initial factual investigation
II. Pleadings
III. Motions (specify)
IV. Affirmative discovery
  1. oral depositions
  2. documents
  3. interrogatories
  4. admissions
  5. other (describe)
V. Disclosure to adversary
  1. preparation of deponents
  2. documents
  3. interrogatories
  4. other (describe)
VI. Settlement
VII. Trial
  1. trial fact and law memos
  2. witness preparation
  3. preparation of cross-examination

Exhibit 1C

Report of Pending Litigation

(To be furnished monthly or when charges exceed $2,500, whichever comes *last*)

I. Professional charges:

	Name	*Hours*	*Rate*	*Total*
1.				
2.				
3.				
4.				

II. Professional services:

1. pleadings	attach copies
2. motions	” ”
3. memos of fact	” ”
4. memos of law	” ”
5. other	describe

III. Disbursements:
1. transportation
2. hotel
3. meals
4. long distance telephone
5. duplicating
6. overtime
7. outside services
   a. investigations
   b. transcripts and reporting
   c. other (describe)
8. other (describe)

IV. Revision of Litigation Preparation Plan? Yes ____ No ____
(If revised, attack copy.)

V. Brief status report:

Exhibit 2

Retainer Discussion Checklist

1. Legal Personnel
   a. Identify, obtain specialty of, and if possible meet all attorneys to be assigned to matter. Satisfied?
   b. All attorneys to know relevant basic client policies in major areas, such as with respect to civil rights, antitrust, and the environment.
   c. Single responsibility—attorney in charge should also be in charge of related internal firm administration and billing.
   d. Preference to be expressed for assignment of senior rather than junior attorneys, and for the same personnel to remain with matter (and other client matters) throughout.
   e. No personnel changes without prior discussion.
2. Strategy
   a. Factual and legal research to be conducted at inception and philosophy of action determined so as to be consistent throughout.
   b. Memorandum to be prepared promptly following development of philosophy, setting forth anticipated tactics including timetable, specific steps, defenses, etc. To be revised in writing as appropriate so as always to be current.
3. Client Participation
   a. Client to receive advance drafts of all significant documents (policy statements, pleadings, memoranda) in sufficient time to be able to participate fully in decisions.
   b. Client routinely to receive copies of all other documents, including letters and internal legal memoranda.
   c. Client to participate in all planning and decisions, including even routine adjournments.
   d. Client to participate in actual conduct of matter to extent client wishes, including handling of negotiations and depositions, arguing of motions, and even trying a case or participating in a trial.
      1. Ordinarily, client staff attorney will participate actively at least in the preparation of all client witnesses who testify.

2. Ordinarily, client nonlegal personnel will handle all paraprofessional assignments (document searches, summaries, factual research, indexing, etc.).

4. Professional Charges
   a. No contingency—fee to be based on time alone.
   b. Most-favored-nation treatment—lowest rate and time charges for each attorney.
   c. Policies as to time charges:
      1. How to calculate travel and weekend time when away from home but no work, or work being done for other clients or matters?
      2. Any maximum per day?
      3. Any minimum per business day away from office? Per entry (1/10 hour, 1/4 hour, 1/2 hour)?
      4. Other (e.g., reduction where attorney inefficient; increase where similar research just completed for another client; charging of training and education time, and of joint time spent traveling or conferring for other clients or purposes)?
   d. Regular (e.g., monthly or as predetermined amounts are reached) statements of time and charges to be furnished, broken down by attorney.
   e. No changes in time rates during handling of matter (or perhaps other discrete period such as one year) without prior discussion.
   f. Policies as to disbursements:
      1. Internal duplication expense?
      2. Travel, meals, entertainment; first-class transportation; family transportation?
      3. Luncheon and dinner expense?
      4. Secretarial and other overtime?
      5. Expenses incurred for joint purposes on behalf of other clients or matters?
      6. Other (e.g., publications, messenger, filing)?
   g. Disbursements to be charged only if required for client effort, not because of other firm or personal priorities (e.g., attorney or secretary working nights or weekends because of other client or bar activities during business day, not chargeable to client).

APPENDIX C

SUMMARY OF THE DISPUTE RESOLUTION CONFERENCE CONCERNING 2,4,5-T, HELD JUNE 4–6, 1979

AMERICAN FARM BUREAU FEDERATION,
PARK RIDGE, ILLINOIS, AUGUST 1979

Table of Contents

Preface

The Dispute Resolution Conference held in Arlington, Virginia, June 4, 5, and 6, 1979, was a first attempt to devise a mechanism by which disputes can be resolved. While recognizing that sociopolitical aspects of controversial issues must be considered in the resolution of disputes, a credible interpretation of the body of available scientific data must be a primary requirement for decision-makers. Without such an interpretation, decision-making would rest solely on value judgments, which vary with individuals and change with time. Accordingly, this conference was deliberately designed to test a mechanism by which the differences among scientists might be resolved. Experience in that aspect of dispute resolution would then provide guidance for future conferences in which sociopolitical aspects could be addressed as well.

In planning the Dispute Resolution Conference, the use of a specific controversial issue was judged to be necessary. The herbicide 2,4,5-T became an obvious choice as a model because of its controversiality and its having been in dispute for many years without resolution. Because the herbicide was also in the RPAR process (Rebuttable Presumption Against Registration—EPA's decision-making procedure for reregistration of pesticides), the conference was specifically timed to occur during the one-month public comment period following EPA's issuance of Position Document Number 3. The first week of June was selected for the conference after discussions with the Deputy Assistant Administrator for Pesticide Programs, because late April or early May seemed most probable for issuance of PD-3. Unfortunately, PD-3 had not yet been issued at the time of the conference and it was not yet available at the time this report went to press. Nevertheless, it is important to make the point that the time for the conference was selected so that it would fit into the sequence of decision-making events in the RPAR process.

Selection of participants for the conference began in February. Since the conference was to be an attempt to resolve disputes on the interpretation of the scientific data available, participants were to be scientists who had conducted research with 2,4,5-T or its contaminant TCDD, or were recognized as being expert in a specific field related to one of the disputatious issues about 2,4,5-T. Individuals who had conducted research on 2,4,5-T and TCDD were identified by a review of the published literature. In an attempt to assure that each of the six workshops had participants who represented different points of view about the issues at controversy, letters were sent in March 1979 to six environmental organizations requesting the identification of scientists who should participate in a discussion of the data. The six organizations were: (1) Environmental Defense Fund, (2) Friends of the Earth, (3) Health Research Group, (4) National Audubon Society, (5) Sierra Club, and (6) John Muir Institute. The National Audubon Society responded to the request by suggesting that the Environmental Clearing House might be able to identify participants. None of the environmental organizations suggested participants for the conference.

The same March letter requested the identification of observers who would attend the conference for the purpose of expressing their concerns about 2,4,5-T, and to monitor the discussions of the workshop participants. On May 2, another letter requesting the identification of observers was sent to the six environmental organizations. The Environmental Defense Fund did identify observers and one did attend the entire conference. Friends of the Earth and Sierra Club declined. The three remaining organizations did not respond. The Rachel Carson Trust, which had not been contacted previously, requested an invitation for an observer. An invitation was sent, and the Rachel Carson Trust was represented throughout the conference.

Observer invitations were also sent to many other interested groups and individuals. A complete list of observers is provided in Appendix C.

The reader will note that the style of the workshop reports is variable. The reason for the variability is that each workshop group adopted a style fitted for the subject matter under discussion. The overview group accepted the style of each workshop, exercising only an editorial responsibility for accuracy of expression and grammatical clarity.

We believe the conference has shown that the differences among scientists in the interpretation of available data can be resolved. But science is only one aspect of dispute resolution. It will be necessary to continue the development of dispute resolution mechanisms that include the sociopolitical aspects of controversial issues. A start has been made, but much more is yet to be done. Scientists who attended the Conference expressed their commitment to dispute resolution. A similar commitment is needed from other elements of society whose interests and concerns become a part of the complete decision-making process. Only then will it be possible to resolve conflict in a manner that is acceptable to the majority of the population.

[*Sections omitted*]

Final Report of 2,4,5-T Dispute Resolution Conference

Introduction

Modern American society is faced with the need to make decisions about a burgeoning number of complex public interest problems, such as whether, how, and under what circumstances to use chemicals, nuclear power, computers, or other technology. The answers to these questions are neither simple nor absolute. They depend on "quality of life" decisions about how one wishes to live on this planet.

One common feature of all of these problems is the need to make decisions about scientific matters, many of which are esoteric for the general public. Traditional science seeks to avoid conclusive decisions until all the necessary data, research results, and opinions have been assembled, evaluated, and judged so that a scientific "consensus" can be reached.

We no longer have the luxury of awaiting a final scientific consensus in this traditional sense. Decisions *must* be made now. There is no other alternative. We either do or do not use a chemical; we either do or do not use nuclear power. To await the final, traditional scientific consensus may mean that the barn door was closed long after the animals escaped. We must find a scientific alternative.

In a free and democratic society, the public ultimately must make the final decisions on quality of life issues. Scientists may debate chemical hazards; legislators may evaluate them; administrative agencies may examine them; courts may adjudicate them. But ultimately the public must decide the critical issues.

Responsible public decision-making in a free and democratic society necessarily requires that the public be furnished the best available information in a form that can be understood and used. Science and scientists thus have an obligation to furnish the public their best possible judgments with regard to scientific issues. Scientists are aware that the public's perception of risk varies with the nature of the hazard. For example, the loss of life in a single airplane crash is perceived differently than an equivalent loss of life from auto accidents on a holiday weekend. Science cannot make judgments for society, but science can provide the information that permits more informed judgments.

To this point, science has not been able to furnish the traditional scientific consensus for decision-makers, because the necessary data for such a traditional consensus are not yet available. The unfortunate alternative has been to furnish scientific conclusions produced in the context of the "adversarial" framework. This means that the advocates of opposing interests each present their data and their conclusions in an effort to "win" for the result they seek. The public recognizes such contentions as adversarial ones. It is unable to evaluate the complex evidence furnished, but has nowhere to turn for independent scientific judgment. To this point, the result has been loss of credibility, disaffection, and growing concern about whether our societal institutions are adequate. We see this in the public dismay at such recent developments as at Three Mile Island and the DC-10 disaster.

The 60 independent scientists who have assembled here from throughout the free world, and the 51 observers who have watched and commented are convinced that decision-makers in our society, and finally the public, will be furnished the best available credible scientific evaluations that it is possible to furnish, even where the traditional scientific consensus is not yet possible because so much remains uncertain and unclear. Scientists have come to participate as scientists, dedicated to science and the service of mankind. They have been able to arrive at consensus with regard to the important and controversial issues that have been addressed in this conference. They believe that their pilot effort in developing an effective dispute resolution mechanism represents a contribution to the process by which our society may come to judgments with regard to scientific matters.

We commend this final report and recommendations of this first scientific dispute resolution conference to decision-makers in our society and the public at large, both for its conclusions concerning the specific scientific problem at hand, and as a model of how the quality of future similar decisions may be enhanced.

Organization of This Conference

The convenors of this conference have long believed that our methods for resolving socioscientific disputes are inadequate, and that new methods of dispute resolution must be developed and tested. Academic intellectual pursuit is not enough. A specific, controversial subject of dispute, which has not responded satisfactorily to existing mechanisms, was chosen as the conference topic. It provides almost the perfect model. It has been in bitter controversy for well over a decade. It has been the subject of recurrent activities by scientific advisory committees, legislatures, regulatory bodies, executive and administrative units, and the courts on state, federal and local levels—more so than any other single, discrete product. Little seems to have escaped repeated trial as a dispute resolution mechanism, to this point with little success in achieving satisfactory resolution.

2,4,5-T first came to active public consciousness as a result of its use as a defoliant in Vietnam. The early concern with regard to its potential harmful effects was colored by the acrimony and bitterness of our Vietnam involvement. In 1969, the discovery of TCDD, a highly toxic contaminant in 2,4,5-T came to the fore. As Vietnam faded into the background, other concerns about carcinogenicity and teratogenicity maintained high public interest. Books, feature news articles, national television specials, and other continuing publicity recently focused public concern on the possibility that 2,4,5-T caused abortions (Alsea, Oregon), and a variety of ills, including cancer, to veterans of the Vietnam conflict.

The scientific dispute resolution mechanism proposed and tested in the past three days was one which would bring together qualified scientists throughout the world who had information or expertise with regard to 2,4,5-T issues. The purpose was to see whether in the presence of each other, after being furnished the plethora of available information, after freely exchanging data and their interpretations of it, and after examining conflicting views, these scientists could arrive at a consensus with respect to the available data. To ensure credibility, this was to be done in the presence of "observers" from parties at interest, from the media, and from the public.

No one was to be compensated for this public interest effort. But at the same time, clearly the expenses involved would be significant. Not only must scientists be brought from all over the world, but accommodations must be furnished for meetings, and provision made for a variety of the other support services without which a scientific meeting would be impossible.

The Research Foundation of the American Farm Bureau Federation, using member and industry contributed funds, provided the financial support for the conference. One of the critical key conditions of accepting financial support at the inception and throughout, has been that there be no effort to influence the conference procedures employed, or the selection of participants. There has been no such influence.

Using this model, invitations were extended to known interested persons or organizations in government, academia, industry, and environment and consumer groups, and to foreign scientists who had been involved in aspects of the relevant scientific inquiries.

It is testimony to the dedication of scientists to the public interest and their conviction that something new is needed, that the response was immediate and constructive. Forty-nine scientists attended from 20 states of this nation, as well as 11 from the nations of New Zealand, Switzerland, Italy, Germany, France, Sweden, and Canada. Many came because an effort was underway to test a new mechanism for resolving important socioscientific problems. In addition to these scientists, 51 observers attended, from government, academia, environmental organizations, industry, the media, and even the USSR.

Conduct of the Conference

Drs. Fred H. Tschirley and Theodore C. Byerly served as conference chairpersons and convenors. Together with Dr. William Upholt, they served as overviewers. The conference assembled Sunday evening, June 3, 1979. It began its formal deliberations with a plenary session of all conferees, participants and observers alike, on Monday morning, June 4. Its work was divided into six key workshops. Appendix B lists the workshop subject matters and participants. Observers are listed in Appendix C.

The workshops met for most of the days on Monday, Tuesday and Wednesday. Further plenary sessions were held on Monday afternoon, primarily for comments by observers, and on Tuesday afternoon for interim reports by the six workshops. At the final plenary session, the general conclusions and recommendations of the workshops were presented. Following this session, the overview group met with workshop chairpersons to prepare the report in final form.

Conclusions

We believe that the Conference had before it the significant data, research results, opinions, and other information presently available with respect to the critical issues regarding the herbicide 2,4,5-T and the contaminant TCDD.

As scientists, we know that there is more work to be done. We know that tomorrow new information may modify our present judgments and conclusions.

As scientists, we would far prefer to be able to wait until tomorrow before we express our scientific views. As members of society, however, we know that we canot wait until then. We have an obligation to let the public know today of our conclusions based on evidence available, so that they may come to decisions for themselves.

We accordingly express these conclusions without reservation, and with the conviction that judgments based upon them must necessarily be the best possible judgments at this time.

General and specific conclusions and recommendations are given in each of the Workshop Reports. For the Conference as a whole, we believe that several conclusions need to be stated because they go beyond the specific issues addressed by the Workshops.

1. Effective mechanisms of dispute resolution are needed. Additional conferences should be held so that appropriate models can be developed for a broad range of issues. Participants in future conferences should be representative of the diverse groups at interest for specific issues.
2. Because of advantages to all society for effective dispute resolution, future conferences to develop models for resolving disputes should be financed by public agencies or by private organizations who are not serving a special interest.
3. The Working Groups of this conference generally used the null hypothesis as the basis for their evaluation of information and thus their conclusions. Regulatory agencies generally use an alternate hypothesis—they assume an effect is real and seek evidence to prove that it is not. The regulatory procedure often results in a conclusion that available evidence does not resolve a substantial question of potential hazard. This difference in approach is important and needs to be addressed.
4. Each of the Working Groups reached consensus judgment, even though wide differences on an issue existed at the beginning of evaluation of relevant information in each group.

We believe that this conference has pointed the way towards a new dispute resolution mechanism which will help society to resolve many of the complex problems with which it is faced.

We are aware of our scientific limitations. We are primarily experts in biology, chemistry, statistics, and economics. Other disciplines must be involved in the fashioning of any long-range dispute resolution mechanism of more general application. Not only would this include scientists from other "hard" disciplines, such as nuclear physics and computer technology, but from the "soft" sciences, including sociology, law, government, political science, and the like.

The Conference has concluded. We call upon those who observe and approve, to convene a further conference to deal with the socio-scientific dispute resolution mechanism as such, rather than with any specific controversial issue.

Science is neither good nor bad. It is how science is used by society that counts. We believe that science can help make this world a better place in which to live. We hope that society will learn how to employ science to its benefit. We as scientists will contribute our valiant effort to that end. We hope that others will join with us in the pursuit of the goal of a better life for all.

[*Sections omitted*]

NOTES

1. Science and Society

1. Milton R. Wessel, *Freedom's Edge: The Computer Threat to Society* (Reading, Mass.: Addison-Wesley, 1974), and Bruce Gilchrist and Milton R. Wessel, *Government Regulation of the Computer Industry* (Montvale, N.J.: AFIPS Press, 1972)—earlier books in this series—each dealt with the impact of computers on society.

2. See NRC, "Abnormal Occurrence Event; Nuclear Accident at Three Mile Island," 44 *Federal Register* 45802 (August 3, 1979); "Taste of Doomsday in Pennsylvania Nuclear Accident," *NYT*, April 1, 1979.

3. See, for example, "Curious Fallout at Three Mile Island," *US News & World Report*, September 24, 1979, p. 55; David Burnham, "Three Mile Island Accident: A Cloud Over Atom Power," *NYT*, September 23–24, 1979; Luther J. Carter, Constance Holden, Eliot Marshall, and Jean L. Marx, "The Crisis at Three Mile Island: Nuclear Risks are Reconsidered," *Science*, April 13, 1979, p. 152; William K. Knoedelseder, Jr. and Ellen Farley, "When Fate Follows Fiction—The 'Syndrome' Fallout," *Washington Post*, March 30, 1979.

4. Tom Alexander, "Time for a Cease-Fire in the Food-Safety Wars," *Fortune*, February 26, 1979, p. 94.

5. Congressman Samuel Devine, *Congressional Record*, September 20, 1977, p. H9718.

6. Tom Alexander, "Time for a Cease-Fire," p. 94.

7. BNA, *Industry and Agriculture*, March 29, 1977, p. M5.

8. Research Institute of America, "Recommendations," August 24, 1979, p. 3.

9. "The Computer Revolution," *C&EN*, August 21, 1978, p. 5.

10. "Thalidomide Again," *US News & World Report*, March 12, 1976.

11. "Patients Irradiated Years Ago Sought Because of Cancer Peril," *NYT*, May 14, 1976; "X-Ray Treatment is Linked to Cancer 35 Years After Use," *WSJ*, May 6, 1976.

12. Edward Ranzal, "City Rejects Order by U.S. to Impose Toll on Bridge," *NYT*, September 6, 1975.

13. Editorial, "A One-Year Test of Dirtier Fuel," *NYT*, August 18, 1979.

14. United States v. LaSalle National Bank, 437 U.S. 298 (1978). See also "Developments in the Law—Corporate Crime," *Harvard Law Review* (1979, 92:1227.

15. Milton Handler, "Antitrust—1978," *ATRR* (December 7, 1978), pp. F-1, F-11; United States v. United States Gypsum Co., 438 U.S. 422 (1978); Furnco Construction Corp. v. Waters, 438 U.S. 567 (1978).

16. *ATRR* (June 28, 1979), p. A-28; *ATRR* (January 18, 1979), p. A-26; Peter M. Gerhart, "Report on the Empirical Case Studies Project," January 15, 1979 (as submitted to NCRALP), p. 20; Connie Winkler, "IBM Throws Trial Course in Doubt," *Computerworld*, July 30, 1979; Connie Winkler, "IBM Trial: Justice's Vietnam?" *Computerworld*, May 21, 1979.

17. United States v. Automobile Manufacturers Association, 1969 Trade Cases ¶72,907 (C.D.Cal.), as modified. See Leonard M. Apcar, "Key Restrictions on Auto Makers Are Struck Down," *WSJ*, July 18, 1979; Letter, Douglas M. Costle to John H. Shenefield, *Legal Times*, October 16, 1978.

18. David Ignatius, "Bell Urges Revamp of Antitrust Policies," *WSJ*, August 15, 1979.

19. "U.S., IBM Urged Again to Seek Suit Settlement," *WSJ*, September 17, 1979; "I.B.M. and U.S. Ready for Talks," *NYT*, September 15, 1979.

20. Philip Handler, "Basic Research in the United States," *Science*, May 4, 1979. pp. 474–75; Ina Spiegel-Rösing and Derek de Solla Price, *Science, Technology, and Society* (Beverly Hills, Calif.: Sage, 1977).

21. E.g., Refuse Act of 1899, 30 Stat. 1152; Pure Food & Drug Act of 1906, 34 Stat. 768; Insecticide Act of 1910, 36 Stat. 331; Caustic Poison Act of 1927, 44 Stat. 1406.

22. 21 USC §348.

23. 87 Stat. 884.

24. 42 USC §§4321-4361.

25. Compare, for example, the editorial, "Policy on Infinitesimals," *NYT*, April 30, 1973, with the editorial, "Tampering With the Delaney Clause," *Washington Post*, February 1, 1973. See also Luther J. Carter, "An Industry Study of TSCA: How to Achieve Credibility?" *Science*, January 19, 1979, p. 247, suggesting that the chemical industry has turned to "a religion called risk/benefit assessment" as a way of avoiding stringent government regulation under TSCA.

26. 15 USC §2601.

27. 16 USC §1531. See also TVA v. Hill, 437 U.S. 153 (1978) (the "snail darter" case.)

28. OMB Guidelines, 44 *Federal Register* 23138 (April 18, 1979). This is the kind of analysis which, in 1972, *Government Regulation* pointed out was totally lacking, and which, in 1974, *Freedom's Edge* urged as the preferred approach.

29. S. 382 (proposed), 96th Congress, 1st Session. See also Joe Sims, "Perspective," *Legal Times*, May 7, 1979.

30. "Needed: A Realistic Environment for Science and Technology," *C&EN*, August 28, 1978, p. 28.

31. "Humanizing the Earth," *Rotarian*, June 1973, p. 15.

32. Galbraith, *Economics and the Public Purpose* (Boston: Houghton Mifflin, 1973). See also Aaron Wildavsky, "No Risk is the Highest Risk of All," *American Scientist*, January–February, 1979, p. 32; William W. Lowrance, *Of Acceptable Risk: Science and the Determination of Safety* (Los Altos, Calif.: Kauffmann, 1976; David Okrent, "Comment on Societal Risk," *Science*, April 25, 1980, p. 372; Joint Hearings, "Risk/Benefit Analysis In The Legislative Process," House Subcommittee on Science, Research and Technology, and Senate Subcommittee on Science, Technology and Space, Ninety-Sixth Congress, First Session (July 24–25, 1979); "Public Willing to Balance Risks and Gains," *CW*, May 28, 1980, p. 71.

2. Our Adversarial Nation

1. "Court Overturns the Murder Convictions of 'Tony Pro' Provenzano, Konigsberg," *WSJ*, June 8, 1979. The reversal itself was reversed on a subsequent appeal. Its implications insofar as process and public perception are concerned, however, remain.

2. Rush Loving Jr., "Unraveling the Riddle of the DC-10," *Fortune*, July 16, 1979, pp. 54, 58.

3. Milton R. Wessel, *Rule of Reason* (Reading, Mass.: Addison-Wesley, 1976), pp. 10–12.

4. Editorial, "Clouds Over Storm King," *NYT*, February 25, 1979. See also William Tucker, "Environmentalism and the Leisure Class," *Harper's*, December, 1977, p. 49.

5. See William H. Simon, "The Ideology of Advocacy: Procedural Justice and Professional Ethics," *Wisconsin Law Review* (1978), 29:30.

6. It is proper that a democratic government should be subject to public influences in certain matters. But it is important also that minority interests be heard and considered, especially in dispute resolution. This is not always the case. For example, agencies such as EPA or FDA will make preliminary decisions and then schedule public hearings. The agency legal staff will consider that its primary responsibility during the hearings is to sustain the preliminary agency decision. Where no active intervener is participating, this may mean that some public concerns are overlooked. One way to deal with this might be by a kind of ombudsman approach, in which a separate participant to the hearing is charged with the responsibility of advancing adversary contentions on behalf of the absentee "publics" involved. However, this poses other problems. For example, in dealing with a police "mistaken identity" problem, even defense counsel conceded that government evaluation of another government agency would be "absolutely impractical," *ABAJ* (October 1979), 65:1470, reports:

> (Defense counsel) charged that after . . . the confessed bandit finally came forward, it was "practically impossible to convince the police that they were wrong. That is a natural result of the adversary system. . . . Unfortunately, the only way to avoid it would be to have another arm created to evaluate the prosecution, but that would be absolutely impractical.

Robert Havely, a Columbia Law School student, offers the suggestion that the White House Office of Science and Technology Policy (OSTP) be encouraged to intervene in important socioscientific disputes between private parties. The procedure might be similar to that followed by the U.S. Justice Department with regard to private antitrust cases. Although OSTP would itself speak only for the Executive Branch and not the public generally, its participation would insure that another important perception is added to the dispute resolution process. I hope Havely will pursue his interest and approach.

7. I use the term "paradigm," in the sense employed by Thomas S. Kuhn, in *Structure of Scientific Revolutions* (Chicago, Ill.: U. of Chicago, 1970). Colleagues at both Stanford and Columbia law schools have sometimes assumed that I have been trying to modify the adversarial litigation "paradigm." I emphasize that I do *not* suggest any such change. That would require a far deeper examination of the whole adversarial system than I have presented. I seek only change in some of the *tactics* employed by adversaries, in their own self-interest.

8. Tom Goldstein, "Some Thoughts by S.E.C. Chief," *NYT*, April 27, 1979, quotes SEC Chairman Harold M. Williams as adding (to Solzhenitsyn):

> A litigious society impairs institutional autonomy and leadership and creates institutional paralysis while litigation winds its laborious way through the procedural maze and delays likely to characterize such a society's judicial system.

9. Dickens, *Bleak House* (New York: Signet, 1964), p. 5.

10. J. Arnould, *Life of Lord Denman* (London: Stevens, 1874), 2:92.

11. Roscoe Pound, "The Causes of Popular Dissatisfaction with the Administration of Justice," *ABA Rep.* (Part 1, 1906), 29:395–417.

12. SCM v. Ind. & Com. Res. Corp., Federal Rules Decisions (N.D.Tex. 1976), 72:110–11. These tactics are equally characteristic of government and criminal cases. In Caleshu v. United States, *USLW* (8th Cir. 1978), 46:2417, a concurring judge wrote:

> Even though the Internal Revenue Service is within its legal rights, I cannot refrain from criticizing the tactics it is using in forcing a taxpayer to litigate his claim several thousand miles from his home after the first suit had been brought by the taxpayer in his home state. In my opinion the practical result may very well be the inability of the taxpayer to get his full day in court because of the cost of defending a suit in Hawaii. This type of legal harassment destroys the confidence of individual taxpayers in the fundamental fairness of the system.

Vincent Bugliosi, the former Los Angeles prosecutor who tried the Charles Manson murder case, put it this way:

> But a trial *is* a game and there's no denying it. It's called the adversary system. Both sides try to maximize every advantage and minimize every disadvantage, which is precisely the same thing that happens on a football field. If we're concerned about pure justice, we may have to minimize this game attitude, but if we do that, what happens to the basic presumption of innocence? What happens to the prosecutor's burden of proving guilt?

"Justice for Sale," *Oui*, November 1976, pp. 71, 108.

13. Charles McCabe, "Clean Up Their Act," *San Francisco Chronicle*, October 31, 1978. See also *Fortune*, July 16, 1979, p. 180.

14. See, e.g., Tom Alexander, "Battered Pillars of the American System: Science," *Fortune*, April 1975, pp. 143, 146; Steven Ebbin and Raphael Kasper, *Citizen Groups and the Nuclear Power Controversy: Uses of Scientific and Technological Information* (Cambridge, Mass.: MIT Press 1974), p. 266.

15. *New Columbia Encyclopedia*, p. 2450 (1975 edition).

16. Editorial, "Truth or Power?" *Science*, October 31, 1975. See also Dr. Boulding's more recent excellent discussion of scientific method, the interaction of science and technology, the benefits and costs of science, and the place of science in society generally. "Science: Our Common Heritage," *Science*, February 22, 1980, p. 831.

17. Letter, *Science*, April 23, 1976, p. 318. James Marshall, "Fact Finding in Law and Science," *ABAJ* (October 1979), 65:1442, writes: "The adversary mentality is not appropriate to science. For that matter, it may be a barrier to conflict resolution and the achievement of consensus necessary in many areas in which lawyers function."

18. F. Pollock and F. W. Maitland, *History of English Law* (Cambridge: Cambridge University Press, 1895), 2:667.

19. Walter Kiechel III, "The Strange Case of Kodak's Lawyers," *Fortune*, May 8, 1978, p. 188. The fact that the case was reversed on appeal and is still not finally concluded as of this writing does not vitiate this conclusion. Indeed, it emphasizes the immediate and dramatic effect of such conduct. An appellate court, removed from the fray, can look at what happened far more dispassionately. See Berkey Photo Inc. v. Eastman Kodak Co., *ATRR*, June 28, 1979 (2d. Cir.), p. E-1.

20. Edward Groth III, "Role of Scientific Expert Examined," *Harvard Law School Record* (February 13, 1976), p. 3. Eugene Rabinowitch, "Back into the Bottle?" *Science and Public Affairs*, April 1973, p. 19, writes:

> In adversary proceedings in which science or one of its applications (such as technology, medicine or psychiatry) are involved, both sides enlist the cooperation of experts—scientists for the prosecution and scientists for the defense, scientists

> for the government and scientists for the opposition. This procedure makes a mockery of science; in fact, it often comes dangerously close to its prostitution.

See, also, Paul F. Rothstein, "Trial by Expert: Accelerating Reality in Modern Courts," *Legal Times*, October 29, 1979. p. 16.

21. Wessel, *Rule of Reason*, p. 12.

22. Stuart A. Feldstein, "How Not to React to a Safety Controversy," *BW*, November 6, 1978, p. 26.

23. Charles Goldsmith, "Firestone Case: A Lesson for Lawyers," *Legal Times*, October 23, 1978, p. 32.

24. Feldstein, "How Not to React," p. 65.

25. Connie Bruck, "How Ford Stalled the Pinto Litigation," *American Lawyer*, June 1979, p. 23. Reginald Stuart, in "Ford Faces New Problems as Pinto Homicide Trial is Due to Begin," *NYT*, January 6, 1980, points up how the ultimate issues in these kinds of cases transcend the immediate ones. The article concludes: "The outcome of the trial is expected to have far-reaching implications on the company's future, although a guilty penalty here would cost it only $10,000 on each count. The outcome could not only influence other pending litigation involving the Pinto, company officials acknowledge, but could also further tarnish the Ford's images in the public's eye. *The consequences on business in general are unknown*" (italics added). A law journal editorial following Ford's acquittal commented: "Ford . . . has emerged . . . victorious. . . . But the company, and industry in general, may well have lost the war. . . . The case . . . focused nationwide attention on . . . criminal prosecution of corporations. It would be naive to think that after all this publicity Ford's acquittal will halt what for many corporate executives has become an established legal threat." *NLJ*, March 24, 1980, p. 14. See also Reginald Stuart, "Ford Won in Pinto Case, But the Memory Will Linger On," *NYT*, March 16, 1980.

26. Wessel, *Rule of Reason*, p. 16. Some of my legal colleagues argue that the success of IBM's litigation tactics is demonstrated by the fact that IBM has successfully delayed the government's effort to punish it for over a decade. But this assumes that IBM will lose—or would have lost had the matter come to trial long ago. It also overlooks the enormous cost of the dispute to IBM in terms of reputation, institutional position, and the constraints which any major litigation imposes—to say nothing of the huge dollar outlay and loss of executive time. Sometimes it is far better to get the matter over with, even if it means a loss, than to perpetuate the misery by continued delay. This certainly has been my experience with most "white collar" clients charged with crime.

27. G. Christian Hill, "Court Orders Mutual of Omaha to Pay Policyholder $2.6 Million; Appeal Slated," *WSJ*, November 17, 1976.

28. Editorial, "The Reserve Scandal," *NYT*, January 10, 1976.

29. See "Public is Sour on Business," *Industry Week*, June 12, 1972, p. 26; "The 'Public Interest' is Widely Ignored in Business and Labor, People Say," *WSJ*, October 26, 1976; *Dow Today*, Commentary, May 14, 1976; "The Mirage of the Rising Right," *Mother Jones*, August 1979, p. 5; "America's Growing Anti-Business Mood," *BW*, June 17, 1972, p. 100. The November 1979 issue of *Mother Jones* is almost entirely devoted to condemnation of corporate society. The cover reads, "The Corporate Crime of the Century."

30. See NCRALP, "Report to the President and the Attorney General," January 22, 1979; "Report of the Special Committee for the Study of Discovery Abuse," *ABA Section of Litigation*, October 1977; Barbara A. Curran, *The Legal Needs of the Public* (Chicago: ABAF, 1977); Eric Schnapper, "The Myth of Legal Ethics," *ABAJ* (February 1978), 64:202; Elizabeth Drew, "The Rule of Lawyers," *New York Times*

Magazine, October 7, 1973, p. 16. See also "Those Cases That Go On and On," *Time*, June 27, 1977, p. 40; Martin Tolchin, "President Says Lawyers Foster Unequal Justice," *NYT*, May 5, 1978; Bernard Weinraub, "Rickover Asserts Lawyers' Tactics Hinder Military," *NYT*, March 31, 1979; Tom Goldstein, "Survey Finds Public Critical of Lawyers," *NYT*, February 11, 1978; Art Buchwald, "Good and Bad Lawyers," *Int. Herald Tribune*, February 21, 1978, and response, "The Good and the Bad," *ABAJ* (April 1978), 64:508; Michael G. Gartner, "Was the Mayflower Filled With Lawyers?" *WSJ*, March 2, 1978; Tuli Kupferberg, "An Insulting Look at Lawyers Through the Ages," *Juris Doctor*, October/November 1978, p. 62; "Group Libel at the Grand Ole Opry," *NLJ*, April 30, 1979, p. 17.

Science has not suffered to the same extent, but its credibility is at issue as well. In an editorial entitled, "The Role of Science in Higher Education," John C. Sawhill, President of New York University (on leave) and Deputy Secretary of Energy, has written:

> As someone who has, for the past 4 years, been president of a major urban university, I am not impressed by recent polls purporting to show that in the American public opinion, science and its practitioners are as hallowed as ever. Scientists may consistently rank higher in public esteem than ministers, architects, lawyers, bankers, and congressmen, but from what I have observed on the college campus and in the classroom, faith in the beneficence of scientific endeavor and the promise of technology has been steadily eroding. *Science*, editorial, October 19, 1979, p. 281.

31. See John R. Emshwiller, "Assessing the Atom—Nuclear Industry Faces Bleak Future as Orders Get Increasingly Scarce," *WSJ*, February 8, 1979; R. A. Brightsen, "The Way to Save Nuclear Power," *Fortune*, September 10, 1979, p. 126; Edmund Faltermayer, "Exorcising the Nightmare of Reactor Meltdowns," *Fortune*, March 12, 1979, p. 82, and accompanying editorial; Donald D. Holt, "The Nuke That Became a Lethal Political Weapon," *Fortune*, January 15, 1979, p. 74; David Burnham, "Inactive Reactors: One Year's Toll of Three Mile Island," *NYT*, March 16, 1980.

3. Legal Responsibility

1. See David H. Berg, "If You Dislike Lawyers, Read This," *NYT*, August 20, 1979; Milton R. Wessel, "Tempering the 'Sport' of Corporate Litigation," *NLJ*, November 6, 1978.

2. See Monroe H. Freedman, *Lawyers' Ethics in an Adversary System* (Indianapolis, Ind.: Bobbs-Merrill, 1975).

3. See Douglas E. Rosenthal, *Lawyer and Client: Who's in Charge?* (New York: Russell Sage Foundation, 1970). Rosenthal argues for participatory decision-making between attorney and client, to replace the more common practice in which attorneys make decisions without client involvement. His logic suggests that he would agree that the client should have the decision power, at least where the matter to be decided does not involve an unlawful or unethical course of conduct, and where the client is a sophisticated commercial organization. This is one of the contentions of my book, *Rule of Reason: A New Approach to Corporate Litigation* (Reading, Mass.: Addison-Wesley, 1976).

4. Milton R. Wessel, "Medical Ethics in Litigation," *Bull. N.Y. Acad. Med.* (1978), 54:808.

5. Potter Stewart, Address, "Professional Ethics for the Business Lawyer: The Morals of the Market Place?" (ABA Section of Corporation, Banking and Business Law, Montreal, August 11, 1975). An NLJ editorial, "The Legal Mind," March 3, 1980,

makes the related point (considered in the discussion which follows) in commenting on the aggressive trial tactics of a former federal trial judge turned trial attorney:

> Has [the former judge] sold out? We don't think so. He was a good judge; he shows signs of being a tough litigator. Loyalty to a cause or client, not consistency of opinion, is the hallmark of an attorney. . . .
>
> If you want to know where a lawyer is "coming from," there is only one question you can ask to find the answer:
>
> "Who is your client?"

6. As reported in Jethro K. Lieberman, *Crisis at the Bar* (New York: Norton, 1978), p. 164*n*.

7. *ATRR* (December 7, 1976), p. A-2. Of course, a trial advocate who pretends that such inconsistent views are his own personally, rather than those of different clients, may end up "hoist on his own petard," with his *own* credibility at issue! See also "A Case of Double Issue Advocacy," *ABAJ* (January 1980), 66:97.

8. *ATRR* (June 7, 1979), p. A-15. A later version of the presentation was published as this book went to press. It is far more restrained. Nevertheless, this discussion of the options for document suppression is similar to including witness abduction as one of the options for eliminating adverse oral testimony. "The Control of Documents," *Antitrust Law Journal* (March 1980) 48:35, 42.

9. "Memoir of a Prosecutor," *Commentary*, October 1976, p. 66.

10. "Justice for Sale," *Oui*, November 1976, pp. 71–72.

11. See the initial draft of the ABA Ethics Code Rewrite Committee, published in *Legal Times*, August 27, 1979, stating (p. 28):

> When acting as advocate, a lawyer should explain in advance the general strategy and assess the prospects of success. The client should be consulted on tactics that might injure or coerce others.
>
> On the other hand, an advocate ordinarily should not be expected to describe trial strategy in detail.

See the ABA's later Discussion Draft of its proposed new Model Rules of Professional Conduct, released at the beginning of 1980, Section 1.4, Comment.

12. See Tom Goldstein, "Law Branches Grow Overseas," *NYT*, September 14, 1979; national law firm surveys published in *NLJ* on August 13, October 1 and October 8, 1979, and in *Legal Times* on September 10, 1979; Daniel J. Cantor, "Law Firms Are Getting Bigger and More Complex," *ABAJ* (February 1978), 64:215. Wessel, *Rule of Reason*, especially pp. 34ff. and chapter 4, discusses a number of aspects of "big" law firm management, practice, and problems.

13. See Irving R. Kaufman, "Ethical Dilemmas Confronting the Practicing Lawyer Today," *Saturday Review*, November 1, 1975; Charles J. Meyers, "On the Devitt Committee Recommendations," *Stanford Lawyer* (Spring/Summer 1979) 14:30; Robert L. Clare, Jr. and Marvin E. Frankel, "Qualifications For Trial Lawyers: A Debate," *New York State Bar Journal* (June 1976) p. 290; Marvin E. Frankel, "Curing Lawyers' Incompetence: *Primum Non Nocere*," *Creighton Law Review* (1977), 10(4):613.

14. See Tom Goldstein, "Attorneys Pondering the Ethics of Their Trade," *NYT*, August 19, 1979; Tom Goldstein, "Obligations of Bar to Clients," *NYT*, September 7, 1979; L. Ray Patterson, "Wanted: A New Code of Professional Responsibility," *ABAJ* (May 1977), 63:639; "Initial Draft of Ethics Code Rewrite Committee," *Legal Times*, August 27, 1979, p. 28; Geoffrey Hazard, *Ethics in the Practice of Law* (New Haven: Yale University Press, 1978); Monroe H. Freedman, *Lawyers Ethics in an Adversary System* (Indianapolis, Ind.: Bobbs-Merrill, 1975); Jethro K. Lieberman, *Crisis at the Bar* (New York: Norton, 1978). The ABA's Discussion Draft has provoked considerable

discussion of the ethical issues involved. See Robert J. Kutak, "Coming: The New Model Rules of Professional Conduct," *ABAJ* (January 1980), 66:46; Linda Greenhouse, "Draft of New Code for Bar is Unveiled," *NYT*, January 7, 1980; Barry M. Hager and Emily Couric, "Ethic Rules Alter Lawyer Role," *Legal Times*, January 28, 1980; Linda Greenhouse, "Lawyers' Group Offers a Revision in Code of Ethics," *NYT*, February 2, 1980, and "The Lawyers Struggle to Uphold Their Own Ethics," *NYT*, February 10, 1980; editorial, "A License to Squeal," *WSJ*, February 11, 1980; "Rewriting the Rules Lawyers Should Live By," *BW*, February 18, 1980, p. 66.

15. Compare In re Primus, 436 U.S. 412 (1978) with Ohralik v. Ohio State Bar Association, 436 U.S. 447 (1978).

16. Michael Knight, "Lawyer Panel Urges Public Control of Legal Ethics," *NYT*, January 17, 1979.

17. ABA Section of Antitrust Law, Tabulation of Responses to Complex Litigation Questionnaire, October 2, 1978 (as submitted to NCRALP), p. 25. See also Peter M. Gerhart, "Report on the Empirical Case Studies Project," January 15, 1979 (as submitted to NCRALP), p. 5.

18. Report of the Special Advisory Panel on Ethical Issues in Complex Antitrust Litigation, December 4, 1978 (as submitted to NCRALP), p. 16. The NCRALP final report said with respect to its Ethics Panel Report:

> The report contains some 10 well-reasoned recommendations. That we do not specifically restate all of them here should in no way be taken as a disagreement with any of them—Report to the President and the Attorney General, January 22, 1979, note 16.

19. Editorial, *NLJ*, October 30, 1978.

20. James Nathan Miller, "Why Crooked Lawyers Go Free," *Readers Digest*, September 1979, p. 47.

21. "A Crisis of Double Standards at the Bar," October 27, 1973, p. 33.

22. See Francis X. Clines, "Association of the Bar Here Is Scored as Dominated by Manhattan Brahmins," *NYT*, October 4, 1973.

23. Report of Ad Hoc Committee on Grievance Procedures (1976), p. 38.

24. Tom Goldstein, "Business and the Law," *NYT*, May 19, 1978.

25. Tom Goldstein, "Bar Unit Shifted in Kuntsler Case," *NYT*, January 30, 1974.

26. Tom Goldstein, "Bar Reassesses Kuntsler Action," *NYT*, February 10, 1974. Undoubtedly in response to this kind of problem, the ABA's guidelines now suggest that all disciplinary proceedings be placed under unified state agencies. "A Change at Hand in Policing Lawyers," *NYT*, February 17, 1980. New York City's attorney disciplinary matters have been transferred from the bar to the courts as of April 1, 1980. Marcia Chambers, "Disciplining of Lawyers Passes From Bar to Court," *NYT*, February 15, 1980.

27. BNA Special Report, "Future Law—Lawyers Confront the 21st Century" (1979), p. 5.

28. See, for example, "The Future: Society," *BW*, September 3, 1979.

29. FDA has developed an interesting new approach in connection with the long-pending aspartame sweetener problem. It and each of the two principal adversaries have named one member of a special Board of Inquiry. The Board consists of a neuroanatomist, a pathologist, and a nutritional biochemist. See, "Three-Member Board of Inquiry Named for Aspartame," *Food Chemical News*, August 27, 1979, p. 1. Joann S. Lublin, "Searle's Battle to Introduce Aspartame, Artificial Sweetener, Enters Key Round," *WSJ*, January 28, 1980, states:

> [The FDA] chose a new procedure, a scientific board of inquiry. The board represents a means "to get away from lawyers and to have scientific matters decided by scientists," explains Neal Singletary, a food scientist with the FDA's Bureau of Foods.

See also, "Decision on Aspartame Due This Year," *Science*, February 22, 1980, p. 856; "Hearings Fail to Resolve Aspartame Controversy," *C&EN*, February 18, 1980, p. 25. "Where Are We Going," *NLJ*, February 11, 1980, refers to the forces compelling changes in the practice of law. It concludes:

> Maybe lawyers, the law and the legal profession will alway be carried by clients and events kicking and fighting into the future.

One commentator wrote to me:

> Why has not the legal profession risen up and said to the world, "There are limits to the adversary process and it should only be used in such-and-such situations"? This question is particularly vital when we consider that our state and national legislatures are dominated by lawyers. Why do the courts allow scientific testimony which is primarily value judgment, not scientific conclusions? I don't know the answer to these questions, but I strongly believe the legal profession has a moral and professional responsibility to speak out on the limits of adversary processes. I don't mean just a mild statement here and there. I mean a shout from the rooftops, through bar associations and individually. Incidentally, I can conceive of nothing that would improve the public image of the legal profession more than doing just this. Private memorandum, November 1, 1979.

30. Joseph E. Leininger and Bruce Gilchrist, *Computers, Society, and Law: The Role of Legal Education* (Montvale, N.J.: AFIPS Press, 1973), p. 25.

4. Corporate Responsibility

1. See Irving Kristol, "Business Ethics and Economic Man," *WSJ*, March 20, 1979. Mr. Kristol writes:

> But lawyers are just about the last class of people one wants to see involved in these matters, since for a lawyer (and rightly so) the ethical is simply identical with the legal. Unfortunately, this is also precisely the point of view most congenial to politicians and bureaucrats, who then proceed to regulate business activity with enthusiasm.
>
> Business ethics, in any civilization, is properly defined by moral and religious traditions, and it is a confession of moral bankruptcy to assert that what the law does not explicitly prohibit is therefore morally permissible.

See also Robert Blair Kaiser, "Conscience Debated As Guide for Modern Business," *NYT*, April 20, 1980.

2. See Dodge v. Ford Motor Company, 204 Mich. 459, 170 N.W. 668 (1919), strongly suggesting that even a liberal-minded court would not sustain the use of corporate assets to accomplish purely social ends. There the court ruled against Henry Ford, whose efforts to reduce the selling price of automobiles had also resulted in a reduction of dividends. Ford had written:

> Business and industry are first and foremost a public service. We are organized to do as much good as we can everywhere for everybody concerned. . . . A reasonable profit is right, but not too much. So, it has been my policy to force the price of the car down as fast as production would permit and give the benefit to the users and the laborers with resulting surprisingly enormous profits to ourselves.

Quoted in J. Irwin Miller, " . . . Blame Not the Socialistic College Professor. . . , *Forbes*, August 1, 1977, p. 48.

3. See "Yale Warns of Divestiture," *CW*, July 12, 1978, reporting:

> Yale University has warned that it may divest its holdings in up to 69 companies with operations in South Africa unless those firms alter their racial policies. . . .
>
> Yale's decision follows similar action at other U.S. universities and colleges, some of which have divested their holdings in companies doing business with South Africa and others that have issued warnings of their intent to do so.
>
> Columbia University trustees, for instance, last month advised 44 such corporations in which it owns $80 million in securities that it will divest itself from those "which, after inquiry, respond in a manner manifesting indifference, through act or omission, to the repressive racial policies in South Africa."

4. Medical Committee for Human Rights v. Securities and Exchange Commission, 432 F. 2d 659, 681 (DC Cir. 1970), case moot, 404 U.S. 403 (1971).

5. Private letter, September 16, 1975.

6. Milton Friedman, *Capitalism and Freedom* (Chicago: University of Chicago Press, 1962), p. 133. See also Milton Friedman, "The Social Responsibility of Business is to Increase its Profits," *New York Times Magazine*, September 13, 1970, p. 32. One senior business executive has written with regard to these kinds of views:

> In a recent *Newsweek* column, Professor Milton Friedman, commenting on business and pollution, argues that business should not really do anything that costs money and which it isn't compelled to do by law or by direct contribution to profit.
>
> Perhaps Professor Friedman can be excused for his naiveté. He is a tenured professor and hasn't had to meet a payroll.

Miller, "Socialistic College Professor," p. 48.

The *New Yorker*, "Notes and Comment," August 20, 1979, attributes views similar to Professor Friedman's to J. Clayburn La Force, dean of the Graduate School of Management of the University of California, Los Angeles. He is quoted as saying that corporate social responsibility is part of "the newest political fad," and, "What makes us think that a businessman knows which social expenditures would improve society in a general sense? He undoubtedly would do what he thought would be beneficial, but, in fact, would probably end up hurting many people." The *New Yorker*, in typical fashion, concludes:

> That is to say, apparently, that the new, philanthropic, socially responsible businessman would manage to make such a botch of his do-good activities that he would achieve social results much like those achieved by his simple-minded predecessors, who thought corporate profits were ends in themselves. Presumably, stability would be preserved on the corporate scene.

7. "Why an Outmoded Ideology Thwarts the New Business Conscience," *Fortune*, October 1970, p. 106. See editorial, *NYT*, November 25, 1979, "American Myopia, Incorporated," suggesting that concentration on short-run profit maximization has lost major American markets to Japan, including the small car market, because Japan concentrates on the long term; Frank A. Weil, "Management's Drag on Productivity," *BW*, December 3, 1979, p. 14, to the same effect.

8. "Corporate Responsibility in the Age of Aquarius," *Bus. Law.* (1970), 26:513, 517.

9. "Rethinking the Language of Business" *NYT*, August 26, 1979. See Courtney C. Brown, *Beyond the Bottom Line* (New York: Macmillan, 1979); Robert W. Ackerman, *The Social Challenge to Business* (Cambridge, Mass.: Harvard University Press), 1975; Robert W. Ackerman and Raymond A. Bauer, *Corporate Social Responsiveness: The Modern Dilemma*, (Reston, Va.: Reston, 1976). See also Donald E. Schwartz, *Commentaries on Corporate Structure and Governance* (Chicago: ALI/ABA, 1980), and the May 1979 issue of the *Hastings Law Journal* (San Francisco, Calif.) which contain

symposia related to the issues of corporate social responsibility and control; "More Corporate Social Responsibility Needed." *ABAJ* (December, 1979), 65:1863. For a thorough and thought-provoking recent review, see David L. Engel, "An Approach to Corporate Social Responsibility," *Stanford Law Review* (November 1979), 32:1. Engel's article points up how differences in definition have confused the debates.

10. Business Roundtable, "The Role and Composition of the Board of Directors of the Large Publicly Owned Corporation" (1978), pp. 11–13. See also "Special Report: The Future of the Corporation—Part II, Control of the Corporation," *Chief Executive*, Autumn 1979, p. 21; John M. Fedders, "No Time For A Low Profile For Big Business," *Legal Times*, February 4, 1980, p. 8.

11. As quoted in editorial, "The Political Role of Business," *C&EN*, October 23, 1978, p. 3.

12. As quoted in Fred T. Allen, "Corporate Morality: Is the Price Too High?" *WSJ*, October 17, 1975.

13. Miller, "Socialistic College Professor," p. 48. See also Ann Crittenden, "Lone Ranger of Corporate Philanthropy," *NYT*, March 15, 1980.

14. "Corporate Responsibility for Consumer Nutrition Education," *Food, Drug, Cos. Law J.* (1977), 32:433, 440. See also "Special Report: The Future of the Corporation—Part I: The Corporation and Society," *Chief Executive* (Spring 1979), pp. 15, 24; Julius E. Johnson, "Initiatives for the Socially Responsible Corporation," Address, Corporation Associates of the American Chemical Society, Washington, D.C., November 6, 1973.

15. "New Business Watchdog Needed," *NYT*, May 25, 1975.

16. Arthur Schlesinger, Jr., "Government, Business, and Morality," *WSJ*, June 1, 1976; Irving Kristol, "Ethics and the Corporation," *WSJ*, April 16, 1975.

17. "The Annual Report 1978: Thick and Innovative," *BW*, April 16, 1979, p. 114.

18. "How Companies React to the Ethics Crisis," *BW*, February 9, 1976, p. 78. Allied Chemical's success in persuading even ardent critics has been extraordinary. See Thomas C. Hayes, "Complying With E.P.A. Rules," *NYT*, January 16, 1980, and editorial, "Praise For An Ex-Polluter," *NYT*, January 28, 1980, concluding: "No one can guarantee that there won't be another major mishap. But Allied Chemical deserves recognition for an impressive corporate turnabout."

19. "Dow Cleans Up Pollution at No Net Cost," *BW*, January 1, 1972, p. 32. Dow's "product stewardship" program has also been widely acclaimed. It seeks to insure that Dow products will be used responsibly by persons and organizations far on down the chain of distribution. See "New Programs Extend Product Responsibility," *CW*, July 4, 1979, p. 53. See also "Capitalizing on Social Change," *BW*, October 29, 1979, p. 105, reporting:

> Costly mistakes, however, may at last be driving home the importance of watching social trends, and recently a number of companies have begun expanding their forecasts well beyond the realm of economics. Some have set up internal departments to predict the future social and political environment in which they will operate, while others are relying on a growing number of consultants who specialize in such crystal-ball gazing. Social predictions even are trickling into strategic plans, and line managers increasingly are called to task for not following them as closely as technical or pricing trends. The ultimate goal is to prevent unexpected social changes from wreaking havoc with profitability. As Robert L. Thaler, a senior vice-president at Security Pacific National Bank, puts it: "If we don't manage social change, change will manage us". . . .
>
> A handful of companies have become downright formal about factoring social and political concerns into strategic plans. Every Wednesday, for example . . . "

5. Responsibility in Dispute Resolution

1. See William M. Carley, "To Pay or Not to Pay—How Exxon Official Agonized Over Making '71 Italian Contribution," *WSJ*, July 14, 1978; Eberhard Faber, "How I Lost Our Great Debate About Corporate Ethics," *Fortune*, November 1976, p. 180; editorial "Why Managers Cheat," *BW*, March 17, 1980, p. 196.

2. See United States v. Irwin, 354 F. 2d 192 (2d Cir. 1965), *cert. denied*, 383 U.S. 967 (1966) (involving the "way of life" making of payments to U.S. Internal Revenue Service Agents). This is just one of many such examples.

3. See "When Businessmen Confess Their Social Sins," *BW*, November 6, 1978; Arieh A. Ullman, "Corporate Social Reporting: Political Interests and Conflicts in Germany," International Institute for Environment and Society, Science Center, Berlin, June 1978; Clark C. Abt, *The Social Audit for Management* (New York: Amacon, 1977); Renaud Gillet, "We Need a Social Balance Sheet," *Chief Executive*, Autumn 1979, p. 36; Meinholf Dierkes and Raymond A. Bauer, *Corporate Social Accounting* (New York: Praeger, 1973); Raymond A. Bauer and Dan H. Fenn, Jr., *The Corporate Social Audit* (New York: Russell Sage Foundation, 1972).

4. See Douglas E. Rosenthal, *Lawyer and Client: Who's in Charge?* (New York: Russell Sage Foundation, 1970.)

5. John P. Lynch III, "The Growth of In-House Counsel," *ABAJ* (September 1979), 65:1403; Doug Lavine, "Corporate Legal Units Moving Up," *NLJ*, October 9, 1978; Nick Galluccio, "The Rise of the Company Lawyer," *Forbes*, September 18, 1978, p. 168; R. K. Valentine, "The Demand for Bigger and Better Law Departments," *ABAJ* (May 1975), 61:642.

6. The bar associations are beginning to publish information regarding management of the substantive legal functions. See the February 1979 Special Issue of the *Business Lawyer* regarding corporate law departments, and the Summer 1979 issue of the ABA publication, *Litigation*, dealing with litigation management. Wessel, *Rule of Reason* is devoted very largely to management of the corporate litigation function. William A. Groening, Jr., retired general counsel of the Dow Chemical Company, is writing a book to be published by McGraw-Hill in 1980. It should be an important contribution to understanding the substantive responsibilities in this area. The working title is "The Modern Corporate Manager: Responsibility and Regulation."

7. Harvard is considering broad changes in its business school training programs, partly designed to fill this need. See "President of Harvard Seeks a Restructuring in the Business School," *NYT*, April 30, 1979; Harvard, "The President's Report," March 1979. Although not specifically referring to legal matters, a lead item in the May 31, 1979, *WSJ* "Business Bulletin," indicates that many of the nation's 1,100 colleges offering degrees in business are reassessing curricula and attitudes—"Also slated to get more classroom attention: the problem of business policies vs. public policies" (p. 1).

8. Marvin H. Zim, "Allied Chemical's $20-Million Ordeal With Kepone," *Fortune*, September 11, 1978, p. 82.

9. "Eastman Kodak Co. Changes Law Firms in Berkey Photo Suit," *WSJ*, March 30, 1978.

10. U.S. Senate, Subcommittee on Executive Reorganization, Special Hearing, March 22, 1966, Appendix.

11. Reginald Stuart, "G.M.'s Image Under Fire in New Type of Lawsuit," *NYT*, April 3, 1978.

12. See editorial, "The Shield Comes Down," *BW*, May 10, 1976, p. 138, concluding:

> The modern corporation is a huge, complex organization. No man can know everything that goes on in each of its factories and warehouses. But a good executive can see to it that lower management reports what is going on and reports honestly. If the chief executive really does not know what is being done in his name, then the company is too big for him to manage.

13. See "Developments in the Law—Corporate Crime," *Harvard Law Review* (1979), 92:1227, and its discussion of the "impossibility" defense.

14. Joseph C. Goulden, "Telex v. IBM," *Juris Doctor* (August/September 1978), p. 28; Becky Barna, "Interim Relief Against IBM Supported by Kennedy," *Computer Decisions* (May 1978), p. 6; W. David Gardner, "From Here to Eternity," *Datamation* (July 1976), p. 51; Larry Lettieri, "Looking for Mr. Goodfriend, or Someone Like Him," *Computer Decisions* (May 1976), p. 58. *Computerworld's* report of the death at 87 of an IBM former senior attorney is a particularly tragic example of this. The attorney had enjoyed a most distinguished legal career, having been the senior partner of one of our nation's great law firms, an appellate court judge, and a leader of the bar. Yet the report ignored most of this, and dealt almost entirely with his alleged trial delaying tactics. "Lawyer for IBM Dies at 87," *Computerworld*, February 11, 1980.

15. David Williams, "Shenefield Hits Katzenbach on Trial 'Delay' Remarks," *Electronic News*, March 6, 1978; Edith Holmes, "U.S. vs. IBM Shows Shambles of Antitrust Law, Senate Told," *Computerworld*, May 9, 1977.

16. *ATRR*, April 27, 1978, p. A-12. See Richard A. Shaffer, "IBM Chairman Estimates Output Boosts, Lashes Out at Justice Department Lawsuit," *WSJ*, April 25, 1978; "Talking Back," *Forbes*, May 29, 1978, p. 116.

17. "IBM Opens Defense," *Electronic News*, May 1, 1978.

18. Tom Goldstein, "Bid to Disqualify a Federal Judge," *NYT*, July 27, 1979.

19. "U.S. Asserts That IBM Didn't Establish Cause to Change Trial Judge," *WSJ*, August 13, 1979; Larry Bodine, "Judge Must Decide if He's Jeckyll or Hyde," *NLJ*, September 10, 1979.

20. "U.S., IBM Urged Again to Seek Suit Settlement," *WSJ*, September 17, 1979; "I.B.M. and U.S., Ready For Talks," *NYT*, September 15, 1979. See also Connie Winkler, "U.S. vs. IBM Feared as Lengthy as World War II," *Computerworld*, October 22, 1979; Arnold H. Lubasch, "IBM Loses Bid to Oust Judge of Antitrust Case," *NYT*, February 26, 1980.

21. Business Roundtable, "Memorandum for the President," March 27, 1978.

22. Steven Ebbin and Raphael Kasper, *Citizen Groups and the Nuclear Power Controversy: Uses of Scientific and Technological Information* (Cambridge, Mass.: MIT Press, 1974).

23. Milton R. Wessel, *Rule of Reason: A New Approach to Corporate Litigation* (Reading, Mass.: Addison-Wesley, 1976), discusses the enormous costs of litigation, and offers suggestions as to how they may be limited and controlled.

24. See, for example, Ralph Nader, Mark Green, and Joel Seligman, *Taming the Giant Corporation* (New York: Norton, 1976); Morton Mintz and Jerry S. Cohen, *Power, Inc.* (New York: Viking, 1976); Leonard Silk and David Vogel, *Ethics and Profits: The Crisis of Confidence in American Business* (New York: Simon and Schuster, 1976); Christopher D. Stone, *Where the Law Ends* (New York: Harper & Row, 1975). See also Christopher D. Stone, "The Corporate Fix," *Center Magazine* (July/August 1976) p. 15; "End of the Directors' Rubber Stamp," and related edi-

torial, *BW*, September 10, 1979; Ralph Nader and Mark Green, "Corporate Democracy," *NYT*, December 28, 1979; "Antibusiness Forces Aim at Corporations," *BW*, Feburary 11, 1980, p. 42. There is even a comic book, "Corporate Crime Comics," dealing with matters such as corporate pollution, unsafe products, tax avoidance and bribery. "Crime in the Suites," *NYT*, February 17, 1980. A number of companies have already been forced to yield substantial control to "outside" directors, including Phillips Petroleum Company, Lockheed Aircraft Corporation, and Gulf Oil Corporation. This appears to be an SEC objective. See "United Merchants Gets Conditions to Stay in Chapter 11 Status," *WSJ*, March 8, 1978.

25. *New Yorker*, June 18, 1979, p. 106. In somewhat related fashion, Thomas A. Murphy, General Motors chairman, has written:

> Freedom, it has been said, is nothing more than an opportunity to discipline ourselves, rather than to be disciplined by others. If we in business want to remain as free as we still are to respond to the desires of our customers, rather than to those of government regulators, we are going to have to fulfill the businessman's first, last, and always responsibility: the responsibility to satisfy those customers today, not tomorrow. American business will remain free from excessive government control only to the extent that we meet this responsibility quickly. It comes down to this: if we want to be allowed to compete in satisfying the customer, we must be willing to compete—"Mr. Murphy Takes the Blame," *NYT*, October 10, 1976.

26. Wessel, *Rule of Reason*, p. 23.

27. Address, "Taking the Positive Approach," *Conference Board*, April 12, 1978.

28. "Honesty and Credibility for Chemical Makers," *C&EN*, July 10, 1978, p. 5.

29. See Morris Janowitz, *The Last Half-Century: Societal Change and Politics in America* (Chicago: University of Chicago Press, 1979). In late October 1979, the U.S. Justice Department's Law Enforcement Assistance Administration (LEAA) released a 324-page study entitled "Illegal Corporate Behavior." It reports such widespread findings of corporate law violation as to suggest that senior executives do in fact have knowledge of what is going on, although they avoid direct participation. See also Leonard Silk and David Engel, *Ethics and Profits;* Eve Pell and Mark Dowie, "Mission Improbable," *Mother Jones*, February/March, 1980, p. 28.

30. See Brian D. Forrow, "The Corporate Law Department Lawyer: Counsel to the Entity," *Business Lawyer* (July 1979), 34:1797. Forrow is general counsel to Allied Chemical Corporation. He has written an excellent analysis of this whole area. One business colleague has commented:

> I don't see what is wrong with having general counsel as a member of senior management and/or the Board. Unless the rest of management and/or the Board defers too heavily to general counsel, he becomes only one vote and one voice. I think that it is hard finding Board members as well educated on proper Board conduct and with as much vested interest in the success of the company as is general counsel. Perhaps there should be a legal audit function conducted by someone other than general counsel, but I think it would be a sacrifice to drive a wedge between counsel and company. Private letter, September 4, 1979.

Another corporate legal colleague observed:

> I am well aware of the risk of possible conflict of interest where an attorney serves a corporation as a director and also as either inside or outside counsel, but I believe this is a matter the corporation should solve for itself. I don't want any outside authority telling a corporation who should or should not be serving on its board or in any other capacity. Once we cross the line and say a lawyer representing a corporation in a socio-scientific matter can't serve on its board, then it's very easy to go the next step and say that any one having any relationship with the corporation can't serve on its board. A university professor could be disqualified from

serving on a corporate board because the corporation has research contracts occasionally with the university and also contributes to it. Then the next step is to mandate who *must* be on the board—a black, an hispanic, a woman, a labor leader, an environmentalist, etc. Private letter, September 3, 1979.

31. See "What the SEC Expects From Corporation Lawyers," *Fortune*, October 23, 1978, p. 143; "Should a Lawyer Be On a Client's Board?" *BW*, November 16, 1974, p. 106; A. A. Sommer, Jr., "Attorneys on Boards: A Growing Concern," *Legal Times*, August 6, 1979, p. 9; Barry M. Hager, "Lawyers on Corporate Boards: Resisting SEC Pressure," *Legal Times*, October 15, 1979, p. 25. SEC Chairman Harold M. Williams delivered an important address covering this and many other current corporate counsel problems, to the 17th Annual Corporate Counsel Institute in Chicago, on October 4, 1979. Key portions of Williams' address are reproduced in *Legal Times*, October 8, 1979, p. 28.

Section 1.12(e) of the initial draft of the ABA Ethics Code Rewrite Committee, phrases the alternatives under consideration as follows:

> A lawyer shall not serve as general counsel of a corporation or other organization of which the lawyer is a director.
>
> Alternative (a): Except upon adequate disclosure to and consent of all persons having an investment interest in the enterprise.
>
> Alternative (b): When doing so would involve serious or recurring risk of conflict between the lawyer's responsibilities as general counsel and those as director. *Legal Times*, August 27, 1979, p. 32.

Section 1.9(f) of the ABA's subsequent Discussion Draft of the proposed Model Rules of Professional Conduct shifts the emphasis somewhat. It reads:

> A lawyer may serve as general counsel to a corporation or other organization of which the lawyer is a director only if:
> (1) There is adequate disclosure to and consent by all persons having an investment interest in the organization; or
> (2) When doing so would not involve serious risk of conflict between the lawyer's responsibilities as general counsel and those as director.

Frederick Andrews, "Management—The First Draft of a New Constitution for Corporations," *NYT*, April 22, 1977, reports the discussions of "a thoughtful assortment of prominent figures in business, government, and the professions, gathered by the American Assembly, a Columbia University affiliate, to discuss and debate 'the ethics of corporate conduct.'" With regard to this issue, their report urged:

> That corporate boards keep separate "the functions of directing the corporation and profering professional counsel"—an urging that a company's legal counsel, investment bankers or similar advisers not serve as board members.
>
> That companies pay "special attention" to "persons representing corporations to all levels of government" and keep them on a short tether.

32. Forrow, "Corporate Law Department," argues that a "general counsel who serves as a director gains a knowledge of and insight into the operations and plans of the corporation which better enables him to render meaningful legal advice" (p. 1817). If the general counsel is given full access to board information, this argument would appear to be somewhat negated.

33. Milton R. Wessel, "Tempering the 'Sport' of Corporate Litigation," *NLJ*, November 6, 1978, p. 19.

34. See Geoffrey Hazard, "Talking to Your Lawyer," *MBA* (May 1978), p. 17. "SEC Wants Comments on Whistleblowing," *ABAJ* (August 1979), 65:1160, reports on the SEC's request for comments on a proposal by Georgetown University's Institute for Public Representation, which "seeks to make explicit an attorney's obligation to the corporation as a whole, rather than just to management." The article quotes the pro-

posal as stating that "in practice, the lawyer too often treats the chief executive officer as his client, without sufficient regard to his or her responsibilities to the board of directors, shareholders, potential investors, and the general public." It quotes Institute Director Charles Halpern as saying that "the corporate attorney serves as advisor to a multi-component 'client' composed of separate constituencies with differing and conflicting interests" and that the ABA Code of Professional Responsibility offers no guide for those situations where the goals of management may conflict with those of the corporation's board of directors or shareholders. Judith Miller, "What a Lawyer Owes His Client," *NYT*, December 17, 1978, states:

> Attorneys say that chief among the ethical questions facing attorneys who advise corporations are the following:
>
> • Is the corporation's attorney responsible to a company's shareholders, management, directors or to all of them?
>
> • If the interests of a company's shareholders, management and directors are in conflict, who should the corporate attorney represent?
>
> • If a corporate counsel accidentally discovers questionable, or potentially illegal, conduct within a corporation, what should he do? Remain silent? Report his suspicions to the board? Resign? Blow the whistle on his clients to the appropriate legal authority?"

See also Geoffrey Hazard, *Ethics in the Practice of Law* (New Haven: Yale University Press, 1978); Stephen Solomon, "The Corporate Lawyer's Dilemma," *Fortune*, November 5, 1979; "SEC Corporate Counsel Proposal Draws Fire," *ABAJ* (January 1980), 66:31.

35. Compare U.S. v. RMI Company, *ATRR* (March 29, 1979), p. A-13, with In re Grand Jury Proceedings, *ATRR* (August 9, 1979), p. A-21. See "Developments in the Law—Corporate Crime: Regulating Corporate Behavior Through Criminal Sanctions," *Harvard Law Review* (1979), 92:1227. Of course this may come too late. One scientist, commenting about this section, writes:

> [T]he book . . . clearly points out that corporate attorneys represent their respective corporations and not the individual managers within them. Legal proceedings in this area demonstrate that this is true. Who then represents the decision-making manager? I believe that many, perhaps most, corporate violations . . . are not deliberate acts of defiance of written regulations, laws or operating permits. Often the rules in question have never been tried in the courts and their exact meaning may be vague and ill-defined. There is a large gray area to deal with. Often, too, choices must be made between physically achievable alternate courses of action, neither of which completely satisfies a strict interpretation of applicable regulations. In addition, in emergency situations one does not first call his lawyer, then act. One acts. Often deep into the night (they always happen at night!), and the next day informs his lawyer of what he has done. Given this situation, what should decision-making managers do? Malpractice insurance might help, but it's of little value where jail sentences are involved.
>
> Well, of course, we all know that prosecutions have not been numerous. They do, however, exist and the potential threat is always there. A mistake in judgement can lead to a jail sentence for a manager who has done what appeared to be best under the circumstances. This problem should be brought out in the open within the corporations and answers to it developed. Private Memorandum, November 1, 1979.

36. This is not to suggest that corporate management will always decide to turn a malefactor over to the authorities for prosecution. There is ordinarily no specific legal obligation to do so. There may be many sound corporate reasons for giving someone a

second chance, extending clemency, etc. See "More Pressure to Prosecute Executive Crime," *BW*, December 18, 1978, p. 104.

37. There are many who believe that a corporate attorney should also have an obligation to the public. "Washington Outlook—Regulation," *BW*, April 9, 1979, p. 129, reports:

> The Securities & Exchange Commission is being asked to extend to corporate lawyers some of the same requirements that it imposed on auditors years ago. Not only would companies have to publicize any change in outside counsel, but lawyers would be obligated to report all suspected violations of securities laws to a company's board. Companies would also have to file written agreements with outside counsel specifying the frequency and nature of the law firm's contact with the board, and the firm's obligation to report illegal corporate conduct.

"When Must a Lawyer Blow the Whistle," *BW*, May 21, 1979, p. 117, states the question as:

> whether lawyers employed or retained by corporations own their allegience solely to management or whether they also have a legal responsibility to the board of directors, shareholders, and even the general public.

Theodore Koskoff, incoming president of the Association of Trial Lawyers of America, presenting his program for the future, "urged the association to play a leading role in the task of holding giant corporations to public accountability." *USLW* (August 14, 1979), 48:2113. See also Tom Goldstein, "Public Trust in Corporations," *NYT*, August 17, 1979.

38. While Assistant Attorney General in charge of the Antitrust Division, John H. Shenefield suggested that this should not be the case. In an address to the ABA Section of Antitrust Law on May 31, 1979, he said:

> Carrying the *Corrugated* instruction [that an effective corporate compliance program might be taken to negate corporate criminal liability] to its logical conclusion would make it virtually impossible to convict a corporation of criminal antitrust violation—a result that is self-evidently nonsensical. The effect of the instruction is to give the corporation an identity independent of the sum of its parts. Thus, regardless of the intent of its agents and employees, the corporation could absolve itself of all liability by virtue of a compliance program. This imbues the corporation with anthropomorphic qualities that clearly are contrary to settled legal principles (p. 13).

39. Louis B. Schwartz, "Tricks and Candor," *Juris Doctor* (Jan., 1977), p. 16. See also Forrow, "Corporate Law Department."

40. Stanley Sporkin, Chief of the SEC's Division of Enforcement, has suggested that there should be "a legal committee that will be responsible for reviewing the legal problems and will be assigned the duty of hiring the corporate lawyer"—"What the SEC Expects," *Fortune*, October 23, 1978, pp. 143, 146. A corporate lawyer has commented that "in many corporations the general counsel reports to a group vice president, administrative vice president or whatever, rather than to the chairman or the president. . . . All this leads me to the point that in advocating reporting to a committee you should not overlook that some corporations need to be told that they make a mistake in having the general counsel report to an intermediate officer and in having lawyers within the corporation report to nonlawyer executives rather than to the general counsel." Private letter, September 3, 1979.

In some corporations, the audit committee performs aspects of the functions of the proposed new legal committee. See also Lawscope, "The Lawyer's Role in Corporate

Management," *ABAJ* (November 1979), 65:1630, for a good general review of the issues in this area. The ABA's 1980 Discussion Draft of the proposed new Model Rules of Professional Conduct treats many of the issues involved in this and other problems discussed in this chapter. It deserves careful study.

41. See "What the SEC Expects," *Fortune*, October 23, 1978, p. 143; "When Must a Lawyer Blow the Whistle," *BW*, May 21, 1979, p. 117; "End of the Directors' Rubber Stamp," *BW*, September 10, 1979, p. 73; Tom Goldstein, "Public Trust in Corporations," *NYT*, August 17, 1979.

42. See *Forrow*, "Corporate Law Department."

43. See Marvin G. Pickholz, "Confronting SEC Pressure: A Need for Legal Audits," *Legal Times*, October 29, 1979, p. 14. See, also, "More Pressure to Prosecute Executive Crime," *BW*, Dec. 18, 1978. This article reports (p. 104):

> Although most companies have long since issued codes of ethics for employees, they possess no comparable guidelines on what to do when the codes are violated. BUSINESS WEEK interviews with corporate executives and law enforcement authorities indicate that, in almost all cases, companies deal with each problem on an individual, catch-as-catch-can basis.

44. See, for example, Nick Galluccio, "The Boss in the Slammer," *Forbes*, February 5, 1979, p. 61; "The Law Closes in on Managers," *BW*, May 10, 1976, p. 110.

45. See "SEC Wants Comments on Whistleblowing," *ABAJ* (August 1979), 65:1160; Judith Miller, "What a Lawyer Owes His Client," *NYT*, December 17, 1978. "Whistle blowing," for purposes of legal obligation concerns, is the opposite side of the coin to "identifying the client." See also "Dissent in the Corporate World: When Does an Employee Have the Right to Speak Out," *Civil Liberties Review* (September/October, 1978), p. 6; Luther J. Carter, "Job Protection for 'Whistle Blowers' Being Tested," *Science*, March 7, 1980, p. 1057. In his October 4, 1979, address, SEC Chairman Williams emphasized that counsel can't sit around and wait for a "trickle" of information. *Legal Times*, October 8, 1979, pp. 28, 29.

46. SEC Chairman Williams indicated his support for this kind of obligation in connection with Foreign Corrupt Practices Act (FCPA) enforcement—the impact of which is by no means limited either to "foreign" or "corrupt" practices—when he advised the ABA Section of Corporation, Banking, and Business Law:

> that one objective of the Commission's much-critized proposal to require management to report on the adequacy of internal accounting controls is to slow the impetus for federal intervention in the private sector by making clear that primary responsibility for the integrity of corporate controls rests on management and the board of directors. The corporate bar, he said, must encourage the corporate self-discipline necessary to avoid such intervention, rather than devise cramped, narrow interpretations of law that hinder the effort toward accountability.
>
> In the chairman's view, counsel's job is to alert management to its responsibilities under the Act, to aid management and the board in structuring the review and decision-making process which an evaluation of controls entails, to help document that effort, and to encourage management to understand the broad corporate accountability concerns which motivated Congress in enacting the accounting provisions.
>
> He urged members of the corporate bar not to take a negative, ultimately self-defeating attitude toward reliable record-keeping. If the profession takes the attitude that the new law should be viewed narrowly and treated as another governmental over-reaction, to be complied with grudgingly, in letter but not in spirit, then the Act will accomplish little except to spawn litigation and harden

the lines between those who urge more pervasive federal control over corporations and those who advocate less. *USLW* (August 28, 1979), 48:2161.

See also "Holding the Line on Auditing Fees," *BW*, October 23, 1978, p. 57, stating that "management and outside auditors soon may have to attest to the reliability of those [FCPA] controls"; "Whom Does the Lawyer Serve?" *BW*, December 17, 1979, p. 104; "A Business Onslaught Scares Off the Sec," *BW*, December 24, 1979, p. 46.

47. See "The End of the Industrial Society," *BW*, September 3, 1979, p. 198, stating: "For while the American technological lead is diminishing, its lead in management is not. American management consulting firms are moving overseas in droves, and they are missionaries for the American system."

48. In a special cover story entitled "Those #*X!!! Lawyers," *Time*, April 10, 1978, p. 66, concludes:

> In the face of all these failures [in our legal system], however, lawyers play ever more important roles. Less and less are they society's servants, more and more the masters of its machinery. That trend is not likely to be halted until clients insist on retaining greater control of the direction of their cases, until citizens give more thought to resolving disputes without plunging into the adversary process, and until voters stop insisting that every perceived wrong be countered with new law and move to reclaim some of the rule-making authority they have consigned to judges and bureaucrats by default.

At least one of my academic legal colleagues is strongly opposed to obligating the attorney to clear legal tactics with the client. In contrast, a business school colleague contends that such a practice would have the additional salutory advantage of helping reduce legal malpractice claims.

6. Responsibility in Science

1. The view that patients should make "medical-risk" decisions, after clear advice regarding anticipated risks and benefits, appears to be taking hold. Jane E. Brody, "Panel on Estrogen Holds Patient's Decision Is Key," *NYT*, September 15, 1979, reports:

> An expert panel convened by the National Institutes of Health concluded today that the decision to use estrogens to treat menopausal symptoms should be based on an individual judgment by the woman after she has been clearly told of known risks and benefits. . . . [T]he group advised that each woman who faces the possibility of treatment should consider the importance to her of obtaining relief from her symptoms and her willingness to face the risks involved.
>
> No one, not even the woman's doctor, can make the decision for her, said the panel's chairman, Dr. Kenneth Ryan, a Harvard University obstetrician and gynecologist. The panel's advice reflects the growing recognition by the medical profession that patients should play an active role in decisions about their care. . . .
>
> The panel said that such benefits [relief of menopausal effects] must be balanced against the risks, especially the risk of endometrial cancer, which is four to eight times more common among users of menopausal estrogens. . . . [T]he group urged that all patients "be given as much information as possible about the evidence for the effectiveness of estrogens in treating specific menopausal conditions and the risks that their use may entail" and that patients "be kept continually informed of new findings as they arise."

Massachusetts has enacted a new "patients rights" law, that guarantees the right "in the case of a patient suffering from any form of breast cancer, to complete information on all alternative treatments which are medically viable." This is reportedly the first such measure in the United States. See "Patients Rights Law Provision Stirs Debate," *ABAJ* (October 1979), 65:1475.

2. Philip Handler, National Academy of Sciences president, stated the following in a May 18, 1979 address at the dedication of Northwestern University's Cancer Center:

> [T]he environmental problems of our day involve risks and benefits that usually accrue to different groups, and costs, risks and benefits that are incommensurable. Costs are reckoned in dollars, benefits in esthetic or material values, risks in human lives. It is for this reason that while risk/benefit analysis can certainly inform the decision maker, his decision must necessarily still turn on a value judgment. The acceptability of a given level of risk remains a political, not a scientific, question. When scientists enter these lists but fail to recognize the boundaries, unspoken idealogies or political beliefs easily becloud seemingly scientific debate. Quoted in *CW*, September 12, 1979, p. 5, "Public Policy Must Rest on Sound Science."

At the SOT's 1976 annual meeting, Russell Peterson, then chairman of the Council on Environmental Quality, said:

> Though scientists have a particular training and expertise which uniquely qualify them to evaluate scientific questions, they possess no special status with respect to social value judgments. . . . It is a social value judgment that, because certain induced tumors transform into malignancies, and because we can seldom predict which tumors will become malignant and which will not, we should regard all tumorigens as possible carcinogens. *PCN* (March 24, 1976), p. 7.

3. "There is no bright line between questions of value and of fact." David L. Bazelon, "Risk and Responsibility," *Science*, July 20, 1979, pp. 277, 278. Another commentator writes:

> In the controversy over nuclear bomb tests, some scientists, called upon by opponents of testing, emphasized the absolute number of radiation-induced bone cancers and leukemias likely to be caused by continued testing in the atmosphere; while other scientists, called upon by advocates of testing, stressed the low number of expected victims, compared to the general incidence of these malignancies. The first group of scientists used the data to claim that continued testing in the atmosphere would be criminal, while the second group used the data to argue that there is no reason to discontinue the tests. Laymen, including legislators, concluded that one cannot trust scientists; some of them say, "Stop tests—they are too dangerous"; others, "Go on, you will not notice the difference." Yet, as scientists, the adversary experts did not disagree on the facts of the situation; they disagreed only on moral conclusions which they derived from these facts—a disagreement in which the judgment of scientists is no more, while no less, valid than that of any other citizen *cognizant of the facts*. Eugene Rabinowitch, "Back into the Bottle?" *Science and Public Affairs* (April 1973), p. 19.

4. William W. Lowrance, in *Of Acceptable Risk: Science and the Determination of Safety* (Los Altos, Calif.: Kaufman, 1976), p. 110, writes:

> In what we have referred to as the "any-man's-land", technical people are presumably as capable as others are, and in many cases more so, because of their breadth of experience and their habit of systematic thought. Not only can they understand the technical details and appreciate the nature of the uncertainties, but from experience they can often provide historical perspective on the problem, anticipate the public's acceptance of the risks fairly accurately, and think of alternatives and consequences that nontechnical people would miss.

Congressman James G. Martin has endorsed this view. In his June 4, 1978, keynote speech to the 38th annual meeting of the Institute of Food Technologists, entitled, "The Role of Scientists in Formulating a National (Rational) Policy for Regulating Carcinogens," he stated (speaking to scientists):

> You must, therefore, get involved. If you do not, it will be left to others to bend the meaning of science so as to fit their advocacy agenda. That is what happened to Soviet biology for 30 years, dominated by the Lysenko doctrine that heredity was controlled by environmental and somatic (whole body) factors rather than by genetics. (p. 4)

Former Stanford geneticist and Nobel Prize winner Joshua Lederberg, now President of The Rockefeller University, seems to suggest a contrary view:

> The able, conscientious men on these [radiation hazard] panels (sometimes including the present author) had no difficulty in finding the common boundary of their knowledge of the hazards of a given dose of radiation. They could make rough estimates of the expected number of deaths and other miseries—but this was all they were competent to do. They should have refused to arrogate the wisdom—which they failed to do—to balance these costs against the anticipated benefits. Instead, the benefits were rarely analyzed, and when dealt with, were stated imprecisely. Indeed, what was demanded of such committees was a policy judgment, cloaked in technical detail. "The Freedom and the Control of Science: Notes From the Ivory Tower," *Southern California Law Review* (1972), 45:609.

5. See the text accompanying chapter 2, note 20, quoting from Edward Groth III, "Role of Scientific Expert Examined," *Harvard Law School Record* (February 13, 1976), p. 3. Rabinowitch, "Back Into the Bottle?" p. 23, writes:

> Scientists, psychiatrists, physicians and technologists should be asked to analyze a problem, and to render their conclusions, without advance presumption as to what point of view they are to defend. If, at a certain point, their conclusions begin to be affected by extra-scientific reasons, they must have sufficient intellectual honesty to state: "Up to this point, I spoke as a scientist; from here on I will speak also as a politically, ethically or ideologically committed citizen."

6. Milton R. Wessel, "Medical Ethics in Litigation," *Bull. N.Y. Acad. Med.* (September 1978), 54:808:

> Medical professionals have a responsibility to themselves and to their profession to act in court in accordance with the same ethical and professional standards as in their day-to-day professional practices. It is not the occupational practitioner's duty to postpone or delay studies so that an attorney may obtain a trial adjournment, to send a reluctant witness with a cold to a hospital so that he need not testify, to withhold data from colleagues so that a witness can be "sandbagged," or to testify other than in the same frank, open, and nonhostile fashion as a patient's question is answered. . . . [T]he assignment of the occupational practitioner is not to solve litigation problems. Instead, it is of course to attend to his own professional obligations and responsibilities. He must not permit himself to be led astray by what appears to be happening elsewhere. And if an attorney suggests a course which seems improper, he should object, seek company review if necessary, and refuse to obey unless and until satisfied.

7. Some of my scientific colleagues doubt that professional societies will ever undertake such an assignment. One writes:

> Several of the societies with which I have been associated have attempted to draw up codes of ethics or otherwise to indicate to their members that certain standards of conduct are necessary and desirable. Invariably these admonitions have fallen

> on deaf ears or have even been resented on the grounds that someone was attempting to interfere with the freedom of the scientist to express his or her views. I do not feel that there is much chance of success along these lines. Private letter, August 29, 1979.

Clearly this has been true in the past. But the timing for action may be ripe today. I am hopeful that aggressive leadership will find substantial support today within many of the scientific societies. I have proposed this course to many audiences of scientists at a number of professional society and other scientist meetings. In every case, the *expressed* consensus was support.

8. Barbara J. Culliton, "Public Participation in Science: Still in Need of Definition," *Science*, April 30, 1976, pp. 451, 452.

9. See also Lawrence K. Altman, "Scientists and Laymen Discuss Ways to Increase Public's Role in the Making of Long-Term Health Policy," *NYT*, April 5, 1976.

10. Culliton, "Public Participation in Science," pp. 451, 452.

11. "New York Academy of Sciences Redefining its Role in Public Policy," *NYT*, November 7, 1976.

12. Peter J. Schuyten, "Scientists and Society's Fears," *NYT*, April 9, 1979.

13. "Identification of Problem Areas for Discussion," SOT Memorandum. This aspect of Golberg's proposal states:

> Erosion of the Toxicologist's prerogative and responsibility for the interpretation of toxicological data has increasingly led to the resolution of scientific disputes by lawyers, judges and other laymen. One consequence has been the perpetration of scientific fallacies, coupled with extensive over-interpretation—that is, exaggeration of toxic hazard, allegedly in the interests of "conservatism", but in reality arising from ignorance of, and lack of experience and perspective in Toxicology. Flowing from these developments the people of the U.S. have had to bear unjustified costs of excessive, unwarranted "protective" measures; they have been denied useful products, and innovative research and development has been stultified. The social and economic consequences in terms of inflation, energy shortage, declining competitive strength of the U.S., adverse balance of payments, and unemployment remain to be fully defined.

14. Philip H. Abelson, "Communicating With the Publics," *Science*, November 5, 1976, p. 565; M. Granger Morgan, "Scientists and Public Policy, *WSJ*, August 26, 1977; Jean L. Marx, "Science and the Press: Communicating With the Public," *Science*, July 19, 1976, p. 136. Under the auspices of the National Academy of Sciences, seventeen scientists from eight countries met in Bellagio, Italy, in June 1976 to consider a broad range of public-policy issues. They presented their conclusions in October, 1976, to a symposium of the National Academy of Sciences held in conjunction with the General Assembly of the International Council of Scientific Unions. One recommendation, as reported in the press, was:

> "Share with the public a sufficient understanding of the risks, technical alternatives and consequences to support wise public policies." The options of those making critical decisions are limited by "the public's sense of priorities and values." Society must often forgo early benefits in the interests of long-term safety or gain. "Unless the public understands the reasons for such decisions, it is difficult for the political leadership, however enlightened," to enable the technical community to make the best choices. Walter Sullivan, "Appeal to Scientists Calls for Activism," *NYT*, October 14, 1976.

DuPont President Edward G. Jefferson has emphasized the need for scientists to assist the public in resolving socioscientific problems. He called for scientists to develop better ways to communicate with the public, to clearly separate value from science, to avoid advocacy as such, and to police themselves to insure responsibility. Address and discussion, CIIT Annual Meeting, March 5, 1980, Research Triangle Park, North Carolina; see "Formula for Toxics Data: Use and Explain," *CW*, March 12, 1980, p. 48.

15. Milton R. Wessel, *Rule of Reason: A New Approach to Corporate Litigation* (Reading, Mass.: Addison-Wesley, 1976), pp. 73–74.

16. Malcolm W. Browne, "How Tiny Chemical Traces are Found," *NYT*, August 14, 1979.

17. Hans Bethe, "Science and the Citizen," *Scientific American*, January 1976, p. 26.

18. Malcolm W. Browne, "How Tiny Chemical Traces are Found," *NYT*, August 14, 1979.

19. Bethe, "Science and the Citizen," p. 26.

20. "The Talk of the Town—Notes and Comment," *New Yorker*, May 21, 1979, p. 25.

21. "Risk and Responsibility," *Science*, July 20, 1979, p. 277.

22. Dick Kirschten, "Can Government Place a Value on Saving a Human Life?" *National Journal*, February 17, 1979, pp. 252, 254.

23. William E. Burrows, "The Cancer Safety Controversy," *New York Times Magazine*, March 25, 1979, pp. 82, 84.

24. "A New Look at the Bottom Line," *C&EN*, June 5, 1978, p. 3. One industry executive who has struggled with this problem has said:

> [T]he dollars vs benefit was not what we were talking about on our cost-benefit curve, but rather we were talking about benefit vs benefit and life vs life. . . . I find this a very difficult concept to get across to people because they don't accept the thought that dollars are the common denominator, and further we in the industry have difficulty expressing the value of our products to people in terms other than dollars." Private letter, October 24, 1975.

25. James P. Lodge, "A Risky Road from Hypothesis to Fact," *BW*, June 21, 1976, pp. 14, 16.

26. *Science*, May 14, 1976, pp. 647, 649.

27. As quoted in "Coming to Grips With Risk," *WSJ*, March 13, 1979.

28. "A Rational Approach to Reducing Cancer Risk," *NYT*, July 9, 1978. The word "calculated" is quoted by me in the text because I suspect Wilson really means "estimated," or even "guesstimated." See also Gio Batta Gori, "The Regulation of Carcinogenic Hazards," *Science*, April 18, 1980, p. 256.

29. "Study Calls Sugar Riskier Than Saccharin," *CW*, August 16, 1978, p. 32.

30. Lowrance ("Acceptable Risk") can be most helpful in this connection. See also Chauncey Starr, "Some Comments on the Public Perception of Personal Risk and Benefit," a paper presented at a Risk vs. Benefit Symposium at the Los Alamos Scientific Laboratory, November, 1971; Wessel, *Rule of Reason*, pp. 24–27. The First Midland Conference of Advances in Chemical Science and Technology, held in October, 1979, represents another new and most interesting approach. Under industry sponsorship but unencumbered by any commercial strings, some 400 scientists and engineers from industry, academe, and government were brought together in what one industry trade journal called a "blockbuster of a program." *CW*, October 31, 1979, p. 29. The purpose of the meeting was to see "how we can get clicking together" on chemical scientific research and development, in the face of the many technical and societal

problems of late-1970s life. A second conference has been called for the fall of 1980. "Air Products Will Host Conference to Bridge Industry-Academia Gap," *CW*, January 23, 1980, p. 35. Carl Friedrich von Weizsäcker, *The Politics of Peril: Economics, Society, and the Prevention of War* (New York: Seabury Press, 1978), argues forcefully that societal demands mandate major changes in the ways scientists approach scientific problems.

31. Philip H. Abelson, editor of *Science*, has already suggested that "a devoted but constructive environmentalist might be asked to serve on the [CIIT] Advisory Committee. There could be other measures. Indeed environmentalists might be consulted for suggestions"—Address, "New Directions in Toxicology," Dedication of the CIIT Laboratory, September 12, 1979. I am not convinced that there is yet enough evidence to indicate that such a step is needed. CIIT is an *industry*-sponsored effort. It does not seek accommodation between conflicting groups, nor to act in any quasi-legislative capacity. I know of no showing that its management is not well-informed regarding the views of nonindustry groups.

32. See the address of then DuPont president Edward Kane upon acceptance of the Palladium Medal Award of the Société de Chemie Industrielle, as excerpted in the editorial, "A Refreshing Voice of Reason from Industry," *CW*, December 24, 1979.

7. Interim Scientific "Consensus-Finding"

1. Alan Anderson, Jr., "Scientist at Large," *NYT Magazine*, November 7, 1976, p. 59, quotes biologist-environmentalist Barry Commoner as saying:

> Most of the major political questions of the day—agriculture, transportation, energy, product safety—have a pretty heavy scientific content and each is connected closely with the economic system. So in dealing with industrial society today, you have to consider a large number of scientific questions.
>
> The second point is that if you ask for a decision on any of these questions, it always gets to be a value judgment. One example I often use is the elm trees and the robins. In the 1960's, when it was shown that spraying elm trees with DDT was killing robins, the public had to decide which it wanted: elm trees or robins. Or take domestic oil reserves. The decision that has to be made is whether we are willing to pay the price of extracting the oil, and how it is going to be paid for. Now those are political questions. If we are going to have a democratic decision, there is simply no way but to supply people with enough information to vote on the issues.

2. Jacobellis v. Ohio, 378 U.S. 184, 197 (1964).

3. Philip Handler, NAS president, has been quoted as suggesting that public policy should not be based on preliminary scientific information. See "Public Policy Must Rest on Sound Science," *CW*, September 12, 1979, p. 5, reporting a portion of Handler's May 18, 1979 address at the dedication of Northwestern University's Cancer Center as follows:

> But surely public policy should not rest on observations so preliminary that they could not find acceptance for publication in an edited scientific journal. . . .
>
> A decade ago it may have been desirable to flag public attention to potential hazard and proceed as if each were a clear and present danger; it is time to return to the ethics and norms of science so that the political process may proceed with greater confidence. The public may wonder at why we don't already know that which appears vital to decision—but science and technology will retain their somewhat diminished place in public esteem only if we steadfastly admit the

> magnitude of our uncertainties and then assert the need for further research. Scientists best serve public policy by living within the ethics of science, not those of politics.

Despite the clear contrary import of his words, I suspect Dr. Handler would agree that public policy must be based upon the best *available* scientific inputs, however preliminary or inadequate they may be. Life proceeds even while scientific investigation continues.

4. For example, two important such reports were released as this manuscript went to press. See "Artificial Sweeteners Seen as Less of Risk than was Thought," *WSJ*, December 21, 1979, and Philip Shabecoff, "U.S. Urged to Lead Drive to Control Fluorocarbons," *NYT*, December 22, 1979.

5. Thus, Steven Ebbin and Raphael Kasper, *Citizen Groups and the Nuclear Power Controversy: Uses of Scientific and Technological Information* (Cambridge, Mass.: MIT Press, 1974), p. 207, quotes NAS President Handler as writing:

> It is essentially impossible to find more than a handful of . . . [nuclear] experts who have not, in relatively recent times, had significant support—either research grants or actual employment—from the U.S. Atomic Energy Commission.

6. Jane Brody, "Consensus Program Praised by Doctors," *NYT*, September 23, 1979. Columbia University President William J. McGill may have had a similar approach in mind in an address to the Guild of Catholic Lawyers of the Archdiocese of New York on September 20, 1977. He said:

> The adversary method for arriving at truth on which our legal procedures are based is, in simple language, not appropriate for arriving at sound public policy on scientific matters. . . . The use of the adversary legal process to control scientific research is likely to lead to serious scientific errors and to badly thought-out policy. . . . How are we to find more responsible ways to make sound public judgments on critical national issues such as the control of energy, science, and technology? . . . [Third, and] finally, the government and the bench should turn more frequently to special commissions constituted from the best and most responsible members of the scientific community in an effort to formulate wise public policy on the protection of the environment, public health, and on all major public safety questions. "McGill Warns vs. Adversary Method," *Science*, October 21, 1977, p. 275.

7. Arthur Kantrowitz is generally given credit for development of the Science Court concept. His proposal is described in "Controlling Technology Democratically," *American Scientist* (1975), 63:505. Two of the best comments on the proposal are James A. Martin, "The Proposed 'Science Court,'" *Michigan Law Review* (1977), 75:1058; and Abraham D. Sofaer, "The Science Court: Unscientific and Unsound," Working Paper No. 1, The Center for Law and Economic Studies, Columbia University School of Law, 1977. There is considerable literature regarding the proposal, much of which is cited in Professors Martin's and (now Federal Judge) Sofaer's articles. Of special relevance to the discussion in this book are: Barry M. Casper, "Technology Policy and Democracy," *Science*, October 1, 1976, p. 29; Arthur Kantrowitz, "The Science Court Experiment," *Jurimetrics Journal* (Summer 1977), p. 332; James A. Martin, "Can Government Deal With Science," *Law Quadrangle News*, Michigan Law School (Spring 1979), 23(3):20; Philip Weinberg, "The 'Science Court' Controversy," *The Record*, Association of the Bar of the City of New York (January/February, 1978), p. 8; "The Science Court," *WSJ*, January 21, 1977; "How the Science Court Would Work," *Dun's Review* (November 1976), p. 78; "The Science Court Experiment: An Interim Report," *Science*, August 20, 1976, p. 653; Philip M.

Boffey, "Science Court: High Officials Back Test of Controversial Concept," *Science*, October 8, 1976, p. 167; "Curbing Ignorance and Arrogance: The Science Court Proposal and Alternatives," *Jurimetrics Journal* (Summer 1979), p. 385.

8. "The Science Court Initiative By The Engineering Society of Detroit," Draft Release, August 3, 1979.

9. For example, see columnist Jack Mabley's favorable quotation of Northwestern University law professor Marshall S. Shapo:

> When people do have a choice, and all the cards are in the table, if the idea of freedom of choice means anything, it means people ought to be able to take risks if they want to.
>
> When people don't know the risks, then that's a case for regulation. If we do know the risks, we ought to be allowed to take them. "How are Products Tested? On Us, of Course," *Chicago Tribune*, August 12, 1979.

The extremely favorable reception to the "Bumpers Bill," introduced by Senator Dale Bumpers in 1979 (S.111), evidences the strong support for a return to legislative decision-making in these matters. The bill would remove the traditional presumption of validity and regularity accorded agency rules and regulations in court. It reverses a generation of trend toward specialized administrative decision-making. Yet the Senate passed the amendment on a 51 to 27 vote, and the ABA House of Delegates, after lengthy debate at its August 1979 Annual Meeting, supported similar legislation on a 146 to 116 vote. *ABAJ* (October 1979), 65:1465.

10. Carson, *Silent Spring*, published by Fawcett Crest, New York, p. 75.

11. Thomas Whiteside, *The Pendulum and the Toxic Cloud* (New Haven: Yale University Press, 1979), and *The Withering Rain* (New York: Dutton, 1971).

12. See Constance Holden, "Agent Orange Furor Continues to Build," *Science*, August 24, 1979, p. 770; "Fallout from Agent Orange Dogs a Herbicide," *BW*, March 24, 1980, p. 114.

13. This may well have been the explanation for attorney misconduct in the *Berkey* v. *Kodak* litigation. See Tom Goldstein, "Easing Stress for Lawyers," *NYT*, December 8, 1978. See also editorial, *NLJ*, October 30, 1978; "The Kodak Legal Morass," *Legal Times*, June 12, 1978; Jeffrey A. Tannenbaum, "Judge's Letter Spurs Probe by Prosecutor of Kodak's Lawyer," *WSJ*, April 11, 1978, and related "Correction and Amplification," *WSJ*, April 12, 1978.

14. One of the most striking examples of this was in the New Jersey Jascalevich curare murder trial, in which the prosecuting attorney was "taunted, insulted and mocked by the defense attorney." David Bird, "Curare Trial Marked by 3-Way Courtroom Strife," *NYT*, April 3, 1978. Bird writes that

> The mounting personal tensions that have embroiled the opposing attorneys and the judge in this unusual trial in Superior Court—sometimes in front of the jury and at times when the jurors are excused—have frequently overshadowed the murder charges themselves.

See also, David Bird, "Jersey Prosecutor Asks Judge in Jascalevich Case to Step Down," *NYT*, April 1, 1978.

15. See Milton R. Wessel, *Rule of Reason: A New Approach to Corporate Litigation* (Reading, Mass.: Addison-Wesley, 1976), pp. 126–27. I must emphasize the critical importance of close communication between client and attorney at all times. If for any reason the client believes it necessary to keep his attorney in the dark, a change of counsel may well be indicated.

16. See Stephen Solomon, "A Businesslike Way to Resolve Legal Disputes,"

Fortune, February 26, 1979, p. 80; Byard G. Nilsson, "A Litigation Settling Experience," *ABAJ* (December, 1979), 65:1818.

17. Raymond W. Fullerton et al., "Final Report on the 2,4,5-T Scientific Workshop," FIFRA, Docket No. 295 (EPA, 1974).

18. Transcript of Conference With Deputy Administrator John Quarles, June 24, 1974, p. 10, as corrected by Order dated July 10, 1974, FIFRA Docket No. 295 (EPA, 1974).

19. 2,4,5-T Dispute Resolution Conference, held June 4–6, 1979 (Reproduced in Appendix C).

20. R. Jeffrey Smith, "Court Reluctantly Upholds EPA on 2,4,5-T Suspension," *Science*, May 11, 1979, p. 602.

21. "2,4,5-T Forum Falls Short of Expectations," *C&EN*, June 11, 1979, p. 7.

22. "2,4,5-T Nearly Problem Free, AFBF Conferees Assert," *PCN*, June 13, 1979, p. 12.

23. Larry Bodine, "Taking on the Deadly Dioxin," *NLJ*, April 2, 1979. See also "Ban Refused on Defoliant," *NLJ*, September 3, 1979, similarly reporting on the Agent Orange litigation, but saying nothing about the conference.

24. Richard Severo, "Agent Orange—A Legacy of Suspicion."

25. Philip Revzin, "Chemical Cloud Still Casts Long Shadow Over Seveso, Italy."

26. Thomas Whiteside, "The Pendulum and the Toxic Cloud," July 25, 1977, p. 34.

27. See "Where We Agree, Report of the National Coal Policy Project, Summary and Synthesis," (Washington, D.C.: CSIS, 1979), pp. 2–3, stating as follows:

> Finding solutions to difficult industry-environmental problems may, in some cases, require that certain conflicting energy, environmental, economic, and social priorities be weighed. Traditionally this task has been accomplished through the adversary process, whereby opposing groups meet at legislative hearings or in the courts to assert their positions. Advocates are forced by the nature of this process to present their case in the starkest terms in order to "win" a favorable decision. This precludes the search for a mutually agreeable outcome. Further, it can lead to additional delays and costs that are in no one's best interest.
>
> The National Coal Policy Project emphasized reaching agreement, where possible, rather than seeking victories. To facilitate this effort, the project adopted a set of negotiating principles known as the "Rule of Reason" . . . Agreement to use these principles helped convince participants that the project could resolve some of the differences constructively, and as it turned out, conducting project meetings in the spirit of the Rule of Reason did facilitate the search for workable solutions to the difficult issues being addressed.

28. "The National Coal Policy Project—Summary" (Washington, D.C.:CSIS, 1977).

29. Tom Alexander, "A Promising Try at Environmental Detente for Coal," *Fortune*, February 13, 1978, p. 95.

30. See Russell E. Train, "The Environment Today," *Science*, July 28, 1978, p. 320; "FTC-Sponsored Workshop Considers Alternative Advertising and Nutrition," *ATRR* (June 29, 1978), p. A-13; William H. Miller, "Movement Toward Environmental Peace," *Industry Week*, February 20, 1978, p. 20; Bill Barich, "Playing Environmental Let's Make a Deal," *Outside*, February 1978, p. 19; "Removing the Rancor From Tough Disputes," *BW*, August 30, 1976, p. 50; "Environmental Consensus," *Resolve* (September 1979), 2(3):1; "Environmental Mediators," *Newsweek*, March 17, 1980, p. 79.

31. Charles Mohr, "Opposing Sides Agree on Ways to Shift to Coal," *NYT*, February 10, 1978.

32. See "Where We Agree, Report of the National Coal Policy Project, Summary and Synthesis," especially the cochairmen's preface at pages xvii–xviii.

33. Luther J. Carter, "Coal: Invoking the 'Rule of Reason' in an Energy-Environment Conflict," *Science*, October 21, 1977, pp. 276, 280.

34. Carol E. Curtis, "Coal Policy Still Needs a Hard Sell," *BW*, February 27, 1978, p. 30.

35. Luther J. Carter, "Sweetness and Light from Industry and Environmentalists on Coal," *Science*, March 3, 1978, pp. 958, 959.

36. See Robert F. Rich and Randy Rydell, "Who Is Making Science Policy?" *The Sciences*, July/August 1979, p. 19:

> The words of Woodrow Wilson seem to be especially appropriate here: "What I fear is a government of experts. God forbid that in a democratic society we should resign the task and give government over to the experts. What are we for if we are to be scientifically taken care of by a small number of gentlemen who are the only men who understand the job? Because if we don't understand the job, then we are not a free people." Harold Laski has similarly written (*Harpers*, December 1930) that "government by experts would . . . mean after time government in the interest of experts."

37. "2,4,5-T Forum Falls Short of Expectations," *C&EN*, June 11, 1979, p. 7.

38. "Sweeteners—Issues and Uncertainties," p. 156 (Washington, D.C.: NAS, 1975).

39. Kenneth R. Hammond and Leonard Adelman, "Science, Values, and Human Judgment," *Science*, October 22, 1976, p. 389. For a criticism of aspects of the approach, see Jordan J. Paust, "Dum-Dum Bullets, Law and 'Objective' Scientific Research: The Need for A Configurative Approach to Decision," *Jurimetrics Journal* (Spring 1978), p. 268.

8. A Plea for Understanding

1. Solicitation Letter, "Who Killed Karen Silkwood?" Karen Silkwood Fund, Youth Project, Washington, D.C., 1979. See Silkwood v. Kerr-McGee Corp., *USLW* (WD, Okla., August 18, 1979, released January 14, 1980), 48:2485.

2. Private letters, October 6, 1975 and October 17, 1977. See Samuel S. Epstein, "The Politics of Cancer," *ABA Barrister* (Winter 1979), p. 11. Alan Anderson Jr., "Scientist at Large," *New York Times Magazine*, November 7, 1976, p. 59, writes with regard to activist Barry Commoner:

> As much as Commoner appeals to those concerned about the environment and to the young, he is openly disliked by industry, mistrusted by labor and regarded with a mixture of envy and outrage by his peers.

3. Letter, "On Eco-Freaks," Douglas Weir, *CW*, February 23, 1977, p. 5.

4. Letter, "Environmentalists vs. Us," Douglas Weir, *CW*, August 29, 1979, p. 5.

5. One scientist advocating environmental causes expressed his reasons for devoting his scientific attention to the newspapers and congressional hearings rather than to the traditional scientific journals, as follows:

> Radical charges against the power structure in the United States required radical methods. . . . The old scientific approach took too long and was too laborious to meet present-day situations. Private memorandum, May 24, 1973.

An attorney representing an environmental organization rejected the scientific-conference approach as "dangerous," "deceptive," and "insidious." He said:

> Industry had the capability to fund studies to generate data, while little organizations had only limited funds. They could not win in the data forum. They can only prevail by using whatever means are available—personal experiences, case histories, and the media. To suggest that they endorse and participate, and endorse the [conference approach], would deprive them of the only real opportunity they have to prove their case. Private memorandum, June 7, 1979.

Research and scientific data produced by industry experts are often rejected as "advocacy." Philip M. Boffey, responding to two legal authorities who had produced a paper which he considered pronuclear, wrote:

> I consider their study "an advocacy brief" in the sense that it was commissioned by, paid for, and distributed by the Atomic Industrial Forum, the trade association of the nuclear industry. The Forum would almost certainly not have commissioned a study by scholars whose findings were apt to undermine its own position. Letters, "Nuclear Power Regulation," *Science*, March 12, 1976, p. 1000.

6. "Against Naderism," *NYT*, April 17, 1974. William R. Barclay, Editor of the Journal of the American Medical Association, has attacked "misleading and inaccurate scientific reports" which have heightened public "anxiety over what substances cause cancer." He appeared to be calling for a similar response when he said:

> Allegations against artificial sweeteners, atomic energy plants, food colorings and preservatives, pharmaceutical products, and industrial chemicals are made almost daily and keep the public in a state of fear that borders on hysteria. "Cancer Reports Criticized," *White Plains Reporter Dispatch*, August 21, 1979.

7. "The Credibility of Corporations," *WSJ*, January 17, 1974. See also Z. D. Bonner, "How Industry Can Regain Public's Trust," *NYT*, September 1, 1974.

8. "The Business of Business," *NYT*, October 6, 1976. See, also, Ron Aaron Eisenberg, "Business Could Try a Touch of Candor," *NYT*, May 13, 1979:

> Business today suffers too frequently from a terminal credibility gap which undercuts its ability to win the public ear. . . . The point is, business might benefit from a touch now and then of candor. . . . It's time for the business community, and those of us in the public relations business to take the lead, to talk "straight" with the American people and Government and the media and ourselves.
>
> There's an important lesson to be learned from the experiences all of us shared recently when a nuclear reactor approached "melt-down" at Three Mile Island. The lesson is candor and moderation and responsibility and balance. The lesson is a basic lesson all of us have learned—nothing really beats the truth.

9. "Brainwashing, Psychiatry, and the Law," *NYT*, May 29, 1976.

10. Willard Gaylin, "What You See Is The Real You," *NYT*, October 7, 1977. Dr. Gaylin is Professor of Clinical Psychiatry at Columbia University, and President of the Institute of Society, Ethics and the Life Sciences. I have quoted somewhat out of context. The context reads:

> I do not care to learn that Hitler's heart was in the right place. A knowledge of the unconscious life of the man may be an adjunct to understanding his behavior. It is *not* a substitute for his behavior in describing him.
>
> The inner man is a fantasy. If it helps you to identify with one, by all means, do so; preserve it, cherish it, embrace it, but do not present it to others for evaluation or consideration, for excuse or exculpation, or, for that matter, for punishment or disapproval.

> Like any fantasy, it serves your purposes alone. It has no standing in the real world which we share with each other. Those character traits, those attitudes, that behavior—that strange and alien stuff sticking out all over you—*that's the real you!*

11. Constance Holden, "Carl Rogers: Giving People Permission to be Themselves," *Science*, October 7, 1977, p. 31.

12. BNA, "White-Collar Justice," April 13, 1976, a lengthy special report on white-collar crime, begins with the paragraph:

> William Pollack is a convicted felon, but he doesn't consider himself a criminal. "I couldn't live with myself if I did," he says emphatically. "I don't consider myself a criminal." Similar examples are legion.

13. See Special Report, "The Future," *BW*, September 3, 1979, p. 167, and the numerous sources it cites; Milton R. Wessel, *Freedom's Edge: The Computer Threat to Society* (Reading, Mass.: Addison-Wesley, 1974).

Index